Proceedings
of the
Steklov Institute of Mathematics

edited by

S. M. Nikol'skiĭ

Number 122 (1973)

Difference methods of solving problems of mathematical physics. II

edited by

N. N. Janenko

Providence, Rhode Island

AMERICAN MATHEMATICAL SOCIETY

1975

Академия Наук
Союза Советских Социалистических Республик

ТРУДЫ

ордена Ленина

МАТЕМАТИЧЕСКОГО ИНСТИТУТА

имени В. А. СТЕКЛОВА

CXXII

Ответственный редактор
академик С. М. НИКОЛЬСКИЙ

РАЗНОСТНЫЕ МЕТОДЫ РЕШЕНИЯ ЗАДАЧ
МАТЕМАТИЧЕСКОЙ ФИЗИКИ. ЧАСТЬ II

Сборник работ
под редакцией
Н. Н. ЯНЕНКО

Издательство ''Наука''

Москва 1973

Translated from the Russian by S. Smith

AMS (MOS) subject classifications (1970). **Primary 35A30, 35A35, 35A40, 65M05, 65N05, 65N20, 65P05, 76G15, 76G20, 76S05, 76T05; Secondary 35L45.**


```
Library of Congress Cataloging in Publication Data
Main entry under title:
Difference methods of solving problems of mathe-
    matical physics.
    (Proceedings of the Steklov Institute of Mathe-
matics ; no. 122)
    Translation of Raznostnye metody resheniia zadach
matematicheskoi fiziki.
    Includes bibliographies.
    1.  Difference equations--Numerical solutions--
Addresses, essays, lectures.  2.  Gas dynamics--
Addresses, essays, lectures.  I.  IAnenko, Nikolai
Nikolaevich.  II.  Series: Akademiia nauk SSSR.
Matematicheskii institut.  Proceedings ; no. 122.
QA1.A413 no.122 [QC20.7.D47]  510'.8s [532'.05'
ISBN 0-8218-3022-8        01515625]  75-20006
```

PROCEEDINGS OF THE STEKLOV INSTITUTE OF MATHEMATICS
IN THE ACADEMY OF SCIENCES OF THE USSR

(Труды математического института им. В. А. Стеклова, т. CXXII, 1973)

TABLE OF CONTENTS

iii

Trudy Mat. Inst. Steklov.
122 (1973)

Proc. Steklov Inst. Math.
122 (1973)

ON SOME QUESTIONS ARISING IN THE NUMERICAL SOLUTION OF PROBLEMS OF POROUS FLOW OF A TWO-PHASE INCOMPRESSIBLE FLUID

UDC 518.517.951

A. N. KONOVALOV

ABSTRACT. In this paper explicit and implicit difference schemes are studied for solving problems of porous flow of a two-phase incompressible fluid in terms of potentials. As a result of the analysis, a new statement of the problem is given in the variables of water saturation and a function of the total flow. The question of computing sources and sinks in problems of porous flow is investigated. Numerical results illustrating various aspects of the work are presented.

Bibliography: 19 items.

§1. Mathematical statement of the problem in terms of the potentials Φ_i

In deriving the equations describing the flow of a multiphase incompressible fluid in a porous medium that take into account the capillary and gravitational forces, it is assumed that at any point of the medium in question the superficial velocity (i.e. volumetric flow rate per unit area) of the ith phase can be represented by the vector

$$\mathbf{u}_i = -k_i \operatorname{grad} \Phi_i. \tag{1.1}$$

Relation (1.1) is a mathematical formulation of Darcy's law, where the following notation is used:

$$\Phi_i = p_i + \rho_i g h,$$

p_i is the pressure in the ith phase, ρ_i is the density of the ith phase, g is the acceleration of gravity and h is the height of the point in question of the medium above a horizontal reference plane.

For definiteness we shall consider the case when the number of immiscible phases in the flow is equal to two. The index $i = 1$ will refer to the phase being displaced, and the index $i = 2$, to the displacing phase.

If the capillary forces are neglected, the phase pressures are equal; otherwise p_1 and p_2 are connected by the relation

$$p_1 - p_2 = p(s). \tag{1.2}$$

The capillary pressure $p(s)$ is a single-valued function of s, the saturation of the displacing phase ($s_2 \equiv s$); in this connection, clearly,

---

AMS (MOS) subject classifications (1970). Primary 76S05, 76T05, 65N05, 65N20.

$$s_1 + s_2 = 1. \tag{1.3}$$

The equations describing the process of porous flow of a two-phase incompressible fluid are now obtained from (1.1)–(1.3) as the laws of conservation of mass of each phase:

$$-ms' \frac{\partial}{\partial t}(\Phi_1 - \Phi_2) = \operatorname{div} k_1 \operatorname{grad} \Phi_1, \tag{1.4}$$

$$ms' \frac{\partial}{\partial t}(\Phi_1 - \Phi_2) = \operatorname{div} k_2 \operatorname{grad} \Phi_2. \tag{1.5}$$

Here m is the porosity of the medium,

$$s' = \frac{\partial s}{\partial p} < 0 \tag{1.6}$$

and $k_i = k f_i(s)/\mu_i$, where k is the absolute permeability of the medium, $f_i(s)$ is the relative permeability to the ith phase and μ_i is the dynamic viscosity of the ith phase.

Equations (1.4) and (1.5) together with boundary conditions for particular physical problems are presented in [1]. In the same place there is also given a numerical method for solving the problem of porous flow of a multiphase incompressible fluid in terms of potentials. The system of equations (1.4), (1.5) does not belong to the so-called systems of Cauchy-Kowalewski type, i.e. it cannot be solved for the derivatives $\partial\Phi_i/\partial t$, $i = 1, 2$. The difficulties caused by this circumstance in the numerical solution of system (1.4), (1.5) were analyzed in detail by us in [2]. Here, however, we confine ourselves to investigating an implicit scheme of the authors of [1] by applying it to the following model system with constant coefficients:

$$\frac{\partial u}{\partial t} - \frac{\partial v}{\partial t} = \gamma_1 \Delta u, \qquad -\frac{\partial u}{\partial t} + \frac{\partial v}{\partial t} = \gamma_2 \Delta v, \tag{1.7}$$

in which γ_1 and γ_2 are positive constants.

Consider the implicit difference approximation of system (1.7)

$$u_{\bar{t}}^{n+1} - v_{\bar{t}}^{n+1} = \gamma_1 \Delta_h u^{n+1}, \qquad -u_{\bar{t}}^{n+1} + v_{\bar{t}}^{n+1} = \gamma_2 \Delta_h v^{n+1}, \tag{1.8}$$

where

$$\tau u_{\bar{t}}^{n+1} = u^{n+1} - u^n, \qquad \tau v_{\bar{t}}^{n+1} = v^{n+1} - v^n, \qquad \Delta_h = \Lambda_{11} + \Lambda_{22}$$

and Λ_{mm} is the usual three-point difference approximation of the second derivative with respect to the corresponding space variable. The index n is the time index, so that $u^n = u_{ij}^n = u(n\tau, x, y)$, $\tau > 0$, and a point with coordinates $x = ih_1$, $y = jh_2$ belongs to the grid under consideration.

Let $\mathbf{w}^{n+1}$ denote the column vector with components u^{n+1} and v^{n+1}. Then (1.8) can be written as follows:

$$A\mathbf{w}^{n+1} = B\mathbf{w}^n, \tag{1.9}$$

where

$$A = \begin{pmatrix} E - \tau\gamma_1\Delta_h & -E \\ -E & E - \tau\gamma_2\Delta_h \end{pmatrix}, \qquad B = \begin{pmatrix} E & -E \\ -E & E \end{pmatrix}$$

and E is the identity operator, so that $E\mathbf{f} = \mathbf{f}$. We note that for any vector $\mathbf{w}^n$ in the class of solutions of system (1.8) we have

$$\gamma_1 \Delta_h u^{n+1} + \gamma_2 \Delta_h v^{n+1} = 0. \tag{1.10}$$

Therefore (1.9) can be rewritten as follows:

$$A^* \mathbf{w}^{n+1} = B\mathbf{w}^n,$$

where

$$A^* = \begin{pmatrix} E - \tau\gamma_1\Delta_h & -E \\ -E + \tau\gamma_1\Delta_h & E \end{pmatrix}.$$

Thus the vector $\mathbf{w}^{n+1}$ is not uniquely determined by (1.9).

We shall supplement the system of difference equations (1.8) with boundary conditions, e.g. boundary conditions of the second kind. We shall see below that similar boundary conditions are in a certain sense natural for a wide class of problems of porous flow theory. It is then possible to regard the difference scheme (1.8) as a system of linear algebraic equations for the unknowns u_{ij}^{n+1} and v_{ij}^{n+1}. The result just obtained implies that the determinant of this system vanishes. This, incidentally, can also be verified by a direct calculation.

The numerical method of the authors of [1], when applied to our model problem, consists in the determination of $\mathbf{w}^{n+1}$ by means of the iterative process

$$\varepsilon_k (\mathbf{w}^{k+1} - \mathbf{w}^k) + A\mathbf{w}^{k+1} = B\mathbf{w}^n. \tag{1.11}$$

The inversion of the operator $(\epsilon_k E + A)$ in (1.11) is realized by means of an alternating direction method. In (1.11)

$$\varepsilon_k = \begin{pmatrix} \varepsilon_{1k} & 0 \\ 0 & \varepsilon_{2k} \end{pmatrix}$$

and the vector $\mathbf{w}^{k+1}$ has the components $(u^{n+1})^{k+1}$ and $(v^{n+1})^{k+1}$ with $(u^{n+1})^0 = u^n$ and $(v^{n+1})^0 = v^n$. It is proposed in [1] that one conduct the iterations with $\epsilon_0 > \epsilon_1 > \epsilon_2 > \ldots$ and so on, the iterations being terminated at the time step when

$$\| \mathbf{w}^{k+1} - \mathbf{w}^k \| < \delta, \tag{1.12}$$

where δ is a preassigned constant and the norm of $\mathbf{w}$ is the maximum of the moduli of its components. It can be stated in advance that the smaller ϵ_k is, i.e. the "closer" system (1.11) is to the original system (1.9), the poorer the condition system (1.11).

A numerical experiment was formulated to determine the singularities of the iterative process (1.11). For simplicity, the one-dimensional system

$$\frac{\partial u}{\partial t} - \frac{\partial v}{\partial t} = \gamma_1 \frac{\partial^2 u}{\partial x^2}, \qquad -\frac{\partial u}{\partial t} + \frac{\partial v}{\partial t} = \gamma_2 \frac{\partial^2 v}{\partial x^2} \tag{1.13}$$

was considered in place of (1.7). Let

$$u(x, 0) = u_0 \cos \pi x, \qquad v(x, 0) = -\frac{\gamma_1}{\gamma_2} u_0 \cos \pi x, \qquad u_0 = \text{const.}$$

Then the solution of (1.13) that satisfies the homogeneous boundary conditions of the second kind at the boundaries of the interval $0 < x < 1$ has the form

$$u(x,\ t) = u_0 \exp\left(-\frac{\gamma_1\gamma_2}{\gamma_1 + \gamma_2}\,\pi^2 t\right)\cos \pi x,$$

$$v(x,\ t) = -\frac{\gamma_1}{\gamma_2}\,u_0 \exp\left(-\frac{\gamma_1\gamma_2}{\gamma_1 + \gamma_2}\,\pi^2 t\right)\cos \pi x. \qquad (1.14)$$

The implicit scheme (1.9) with operator $A = A_1$, where

$$A_1 = \begin{pmatrix} E - \tau\gamma_1\Lambda_{11} & -E \\ -E & E - \tau\gamma_2\Lambda_{11} \end{pmatrix},$$

and the same operator B as before was taken as an approximation of system (1.13). The problem

$$A_1 \mathbf{w}^{n+1} = B\mathbf{w}^n$$

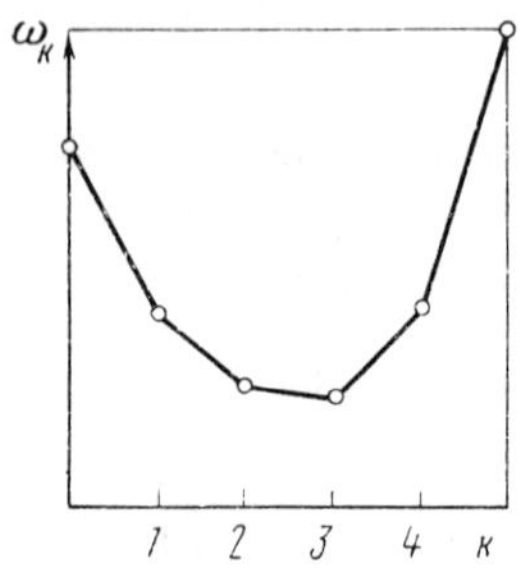

FIGURE 1

was solved by means of the iterative scheme (1.11). A typical result of the calculation is presented in Figure 1. Here the abscissa k is the number of iterations while the ordinate ω_k is defined by the equality

$$\omega_k = \max_i \|\mathbf{w}_i^{n+1} - \mathbf{w}_i^k\|,$$

in which $\mathbf{w}_i^{n+1}$ is calculated from (1.14) while $\mathbf{w}_i^k = (\mathbf{w}_i^{n+1})^k$ is calculated by means of the iterative difference scheme (1.11). The behavior of the curve ω_k could have been predicted in advance: as long as ϵ_k is sufficiently large, the residual is decreased (there also exist calculations of a similar kind in which it is increased); then the point is reached when the smallness of the determinant of system (1.11), i.e. the poorness of the condition of this system, commences to play a decisive role.

Thus the choice of the iterative parameters ϵ_k plays a decisive role in the solution of a degenerate system (1.9) by means of the iterative process (1.11). Until the problem of minimizing ω_k through a suitable choice of the sequence ϵ_k is solved, the iterative method (1.11) of solving system (1.9) cannot be regarded as satisfactory.

§2. Mathematical statement of the problem in terms of s and ψ

Below we consider another statement of the problem of porous flow of a two-phase incompressible fluid. This new statement is justified by the fact that one does not have to deal with ill-conditioned systems of algebraic equations of type (1.11) in a numerical realization of it.

From (1.4) and (1.5) it follows that

$$\operatorname{div} \mathbf{q} = 0, \qquad (2.1)$$

where

$$\mathbf{q} = -(\mathbf{u}_1 + \mathbf{u}_2)$$

or, in more detail, $\mathbf{q} = (q_x, q_y)$ and

$$q_x = k_1 \frac{\partial p_1}{\partial x} + k_2 \frac{\partial p_2}{\partial x} + g \frac{\partial h}{\partial x} (k_1 \rho_1 + k_2 \rho_2),$$

$$q_y = k_1 \frac{\partial p_1}{\partial y} + k_2 \frac{\partial p_2}{\partial y} + g \frac{\partial h}{\partial x} (k_1 \rho_1 + k_2 \rho_2). \tag{2.2}$$

Relation (2.1) permits us to introduce a stream function ψ:

$$q_x = \frac{\partial \psi}{\partial y}, \qquad q_y = -\frac{\partial \psi}{\partial x}. \tag{2.3}$$

For the function ψ introduced in (2.3), relation (2.1) is satisfied identically. We now differentiate (1.2) with respect to x and y:

$$\frac{\partial p_1}{\partial x} = \frac{\partial p_2}{\partial x} + p' \frac{\partial s}{\partial x}, \qquad \frac{\partial p_1}{\partial y} = \frac{\partial p_2}{\partial y} + p' \frac{\partial s}{\partial y}. \tag{2.4}$$

From (2.2)–(2.4) we get

$$\frac{\partial p_2}{\partial x} = \frac{1}{k_1 + k_2} \left[\frac{\partial \psi}{\partial y} - g \frac{\partial h}{\partial x} (k_1 \rho_1 + k_2 \rho_2) - k_1 p' \frac{\partial s}{\partial x} \right],$$

$$\frac{\partial p_2}{\partial y} = \frac{1}{k_1 + k_2} \left[-\frac{\partial \psi}{\partial x} - g \frac{\partial h}{\partial y} (k_1 \rho_1 + k_2 \rho_2) - k_1 p' \frac{\partial s}{\partial y} \right]. \tag{2.5}$$

Substituting (2.5) in (1.5), we obtain the equation needed by us for the saturation s of the displacing phase:

$$m \frac{\partial s}{\partial t} = \frac{\partial}{\partial x} \left[c \left(a_x + \frac{\partial \psi}{\partial y} - k_1 p' \frac{\partial s}{\partial x} \right) \right] + \frac{\partial}{\partial y} \left[c \left(a_y - \frac{\partial \psi}{\partial x} - k_1 p' \frac{\partial s}{\partial y} \right) \right]. \tag{2.6}$$

Here

$$c = \frac{k_2}{k_1 + k_2}, \qquad a_x = k_1 (\rho_2 - \rho_1) g \frac{\partial h}{\partial x}, \qquad a_y = k_1 (\rho_2 - \rho_1) g \frac{\partial h}{\partial y}.$$

Inasmuch as $c > 0$ and $p' = dp/ds < 0$, equation (2.6) is of parabolic type.

We next derive an equation for ψ. Differentiating the first (second) of relations (2.5) with respect to y (with respect to x) and equating the second-order mixed derivatives of p_2, we get

$$\Delta \psi = F, \tag{2.7}$$

where $F = F_1 + F_2$ and

$$F_1 = \frac{k_1' + k_2'}{k_1 + k_2} \left(\frac{\partial s}{\partial x} \frac{\partial \psi}{\partial x} + \frac{\partial s}{\partial y} \frac{\partial \psi}{\partial y} \right),$$

$$F_2 = \frac{k_1' k_2 - k_2' k_1}{k_1 + k_2} g (\rho_2 - \rho_1) \left(\frac{\partial h}{\partial y} \frac{\partial s}{\partial x} - \frac{\partial h}{\partial x} \frac{\partial s}{\partial y} \right).$$

Equations (2.6) and (2.7) will be our initial equations in a numerical solution of problems of porous flow of a two-phase incompressible fluid.

Let us now consider the boundary conditions for s and ψ. Inasmuch as a detailed investigation was made in [3] of the boundary conditions for problems of porous flow of a two-phase incompressible fluid in which capillary forces are taken into account, we confine ourselves here to an elementary concrete physical problem.

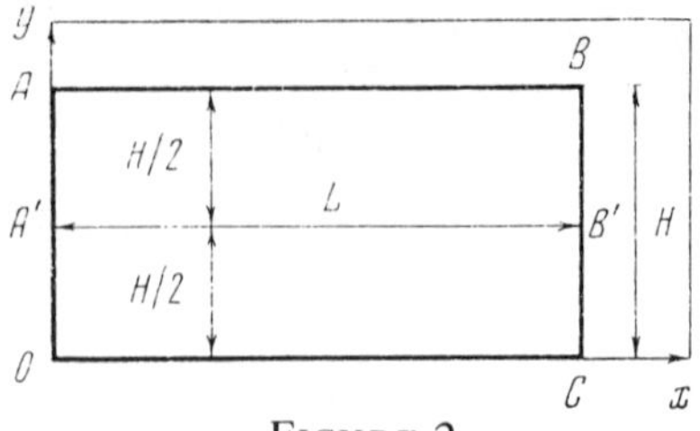

FIGURE 2

PROBLEM 1. Suppose given an oil-bearing stratum $OABC$ with a known initial distribution of water saturation (Figure 2). The upper and lower boundaries AB and OC of the stratum are assumed to be impermeable to water. Commencing at the moment of time $t = 0$, water (the displacing phase) is injected into the stratum through the inflow face AO. As to the face BC, it can be assumed that only oil (the displaced phase) flows through BC for $0 \leqslant t \leqslant t_0$ (t_0 is the so-called breakthrough time), while a mixture of oil and water flows through BC for $t > t_0$. Or it can be assumed that a mixture of the two phases flows through BC from the very beginning, in which case it is necessary to indicate in addition the fraction of each phase in the stream.

We now give a mathematical formulation of the boundary conditions of Problem 1. The impermeability means that there is no flow through the faces AB and OC, i.e.

$$(u_1)_y = -k_1\left(\frac{\partial p_1}{\partial y} + \rho_1 g \frac{\partial h}{\partial y}\right) = 0, \tag{2.8}$$

$$(u_2)_y = -k_2\left(\frac{\partial p_2}{\partial y} + \rho_2 g \frac{\partial h}{\partial y}\right) = 0, \tag{2.9}$$

where $(u_i)_y$ is the component of the vector $\mathbf{u}_i$ in the direction perpendicular to the faces AB and OC. Only water flows through the inflow face (there is no flow of oil), which means

$$(u_1)_x = -k_1\left(\frac{\partial p_1}{\partial x} + \rho_1 g \frac{\partial h}{\partial x}\right) = 0, \tag{2.10}$$

$$(u_2)_x = -k_2\left(\frac{\partial p_2}{\partial x} + \rho_2 g \frac{\partial h}{\partial x}\right) = Q_0. \tag{2.11}$$

If only oil flows through the outflow face, then

$$(u_1)_x = -k_1\left(\frac{\partial p_1}{\partial x} + \rho_1 g \frac{\partial h}{\partial x}\right) = Q_0, \tag{2.12}$$

$$(u_2)_x = -k_2\left(\frac{\partial p_2}{\partial x} + \rho_2 g \frac{\partial h}{\partial x}\right) = 0. \tag{2.13}$$

For the problem under consideration it is necessary to put $h(x, y) = y$. The boundary conditions for the water saturation are then obtained from (2.8)–(2.13) by taking (2.4) into account. On the impermeable faces AB and OC we have

$$p' \frac{\partial s}{\partial y} = g\,(\rho_2 - \rho_1); \tag{2.14}$$

on the inflow face OA

$$k_2 p' \frac{\partial s}{\partial x} = Q_0, \tag{2.15}$$

and on the outflow face BC

$$k_1 p' \frac{\partial s}{\partial x} = -Q_0. \qquad (2.16)$$

Inasmuch as the initial water saturation $s(x, y, 0) = s_0(x, y)$ is assumed to be known, the statement of the problem for s is complete.

Now as to the boundary conditions for ψ, by definition,

$$q_x = \frac{\partial \psi}{\partial y}, \qquad q_y = -\frac{\partial \psi}{\partial x}.$$

On the inflow face OA we have

$$q_x = -(u_1)_x - (u_2)_x = -Q_0$$

by virtue of (2.10) and (2.11), and hence

$$\psi(y)|_{OA} = \psi(O) - yQ_0,$$

where $\psi(O)$ is the value of ψ at the point O. Analogously, on the outflow face BC

$$q_x = -(u_1)_x - (u_2)_x = -Q_0$$

and hence

$$\psi(y)|_{BC} = \psi(C) - yQ_0. \qquad (2.18)$$

Here $\psi(C)$ is the value of ψ at C. On the lower boundary OC of the stratum we have

$$q_y = -(u_1)_y - (u_2)_y = 0$$

and hence

$$\frac{\partial \psi}{\partial x}\bigg|_{OC} = 0.$$

In order to satisfy this condition we must put

$$\psi(O) = \psi(C) = \psi_0 = \text{const}, \qquad \psi(x)|_{OC} = \psi_0. \qquad (2.19)$$

Then on the upper boundary AB of the stratum we get

$$\psi(x)|_{AB} = \psi_0 - HQ_0, \qquad (2.20)$$

where $OA = BC = H$. The statement of the problem for ψ is also complete.

If both of the phases flow through BC, the fraction of each of them in the flowing stream can be determined by various methods. One of the most satisfactory methods is the assumption that the fraction of each phase in the flowing stream is proportional to the mobility of this phase, i.e. $Q_0 = Q_1 + Q_2$ and

$$\frac{Q_2}{Q_1} = \frac{f_2(s)\,\mu_1}{f_1(s)\,\mu_2}. \qquad (2.21)$$

Inasmuch as a calculation of the capillary forces assumes that (1.2) holds, (2.21) is equivalent to the fact that

$$\frac{\partial s}{\partial x} = 0. \qquad (2.22)$$

The boundary condition for ψ is not affected by this assumption, since it involves only the total flow through BC and for the given total flow it remains unchanged.

The following algorithm for the solution of the s, ψ statement of Problem 1 can now be proposed.

a) We determine ψ^{n+1}, i.e. the approximation of ψ at the $(n + 1)$th time step, from equation (2.7), the boundary conditions for ψ and the known approximation s^n of the distribution of water saturation at the nth time step. We have a Dirichlet problem for the determination of ψ, so that it is expedient to apply one of the rapidly converging iterative methods of alternating directions (see, for example, [4]).

b) The approximation s^{n+1} of the water saturation at the $(n + 1)$th time step is determined from equation (2.6), the boundary conditions (2.14)–(2.16) or (2.14), (2.15), (2.22) and the known quantities ψ^{n+1} and s^n. Here again it is expedient to apply one of the alternating direction methods.

c) Knowing s^{n+1} and ψ^{n+1}, we find p_2^{n+1} to within an arbitrary additive function of time by integrating its total differential:

$$p_2^{n+1} = \int \left(\frac{\partial p_2}{\partial x}\right)^{n+1} dx + \left(\frac{\partial p_2}{\partial y}\right)^{n+1} dy + f(t). \tag{2.23}$$

The derivatives of p_2 with respect to x and y in (2.23) are given by (2.5), while the expression

$$\frac{\partial p_2}{\partial x} dx + \frac{\partial p_2}{\partial y} dy$$

is a total differential by virtue of the method of deriving equation (2.7).

d) We determine p_1^{n+1} from the equation $p_1^{n+1} = p_2^{n+1} + p^{n+1}$, in which the quantity $p^{n+1} = p(s^{n+1})$ is uniquely determined by the known quantity s^{n+1}.

Having calculated the approximations of ψ, s, p_2 and p_1 at the $(n + 1)$th time step, we proceed to calculate these approximations at the $(n + 2)$th time step, and so on.

We conclude this section with some words on other mathematical statements of the problem of porous flow of a two-phase incompressible fluid that take the capillary forces into account. In [1], besides the statement of the indicated problem in terms of the potentials Φ_i, there is presented another, in our view more felicitous, statement of the problem in terms of the variables P and R:

$$\operatorname{div} M \operatorname{grad} R + \operatorname{div} N \operatorname{grad} P = -4ms' \frac{\partial R}{\partial t},$$
$$\operatorname{div} N \operatorname{grad} R + \operatorname{div} M \operatorname{grad} P = 0, \tag{2.24}$$

where

$$P = \frac{1}{2}(\Phi_1 + \Phi_2), \qquad R = \frac{1}{2}(\Phi_1 - \Phi_2),$$
$$M = k_1 + k_2, \qquad N = k_1 - k_2. \tag{2.25}$$

We note that it would also be possible to solve Problem 1 for P and R. For R, as for s, we would have boundary conditions of the second kind. As for P, in place of a Dirichlet problem, which was formulated by us for ψ, we would have a Neumann problem. In this sense the s, ψ statement of Problem 1 is preferable.

As follows from (2.25), the P, R statement assumes that the boundary conditions for the potentials Φ_1 and Φ_2 are of the same kind. The statement of the problem of porous flow of a two-phase incompressible fluid in terms of the variables s and p_i that was proposed in [3] and takes into account the capillary and gravitational forces permits us to remove this restriction. This statement is ideologically close to (2.24), (2.25). One of the potentials in the expression for P here is eliminated with the use of (1.2), while an equation for s is obtained from the first of equations (2.24) by replacing the derivatives of R by derivatives of s. However—and this is very important—the s, p_i statement, in contrast to the P, R statement, is applicable under any boundary conditions.

§3. Numerical realization of the boundary conditions in porous flow problems

Before passing to an analysis of the boundary conditions, we recall that $p(s)$, $f_1(s)$ and $f_2(s)$ are experimentally measurable functions of the saturation s of the displacing phase. More precisely (see, for example, [5]),

$$p(s) = \frac{I(s)\, \alpha \sqrt{m/k}}{\cos \theta},$$

where $I(s)$ is Leverett's dimensionless function, α is the interfacial tension and θ is the contact angle.

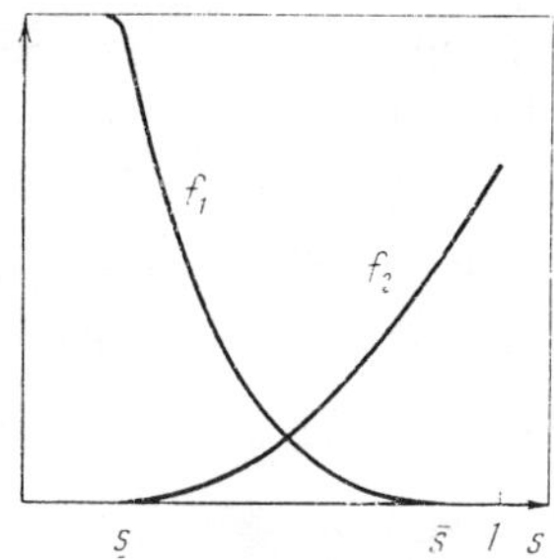

FIGURE 3

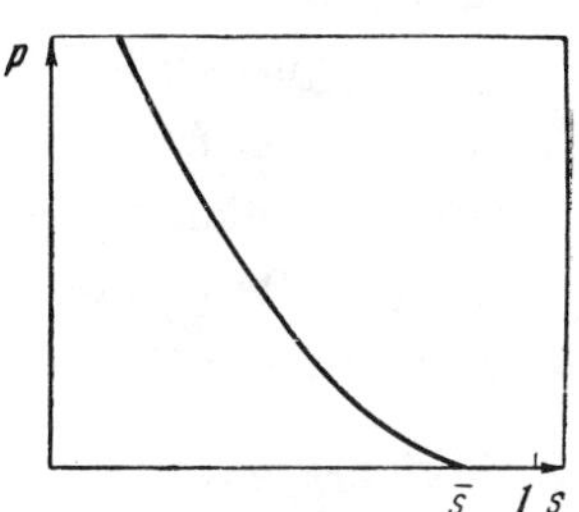

FIGURE 4

As to $f_1(s)$ and $f_2(s)$, a typical relative permeability vs. saturation relation for a water-oil-rock system is shown in Figure 3. A characteristic feature here is the existence of two limiting values $\underline{s}$ and $\bar{s}$ of the saturation of the displacing phase, below and above which the corresponding relative permeability vanishes. It follows from Darcy's law that the displacing phase is stationary for $s < \underline{s}$, while the displaced phase is stationary for $s > \bar{s}$. A typical capillary pressure vs. saturation relation is shown in Figure 4, which, like Figure 3, is taken from [5]. A characteristic feature here is that the capillary effects are apparent only for $s < \bar{s}$ and can be neglected for $s > \bar{s}$. We note that since we are primarily interested in the process of displacing oil by water and since the water saturation in the stratum rises during this process, the curve in Figure 4 can be regarded as the so-called imbibition curve.

We now pass directly to an analysis of the boundary conditions in porous flow problems. In this connection, we will dwell on the s, ψ statement and confine our attention to the boundary conditions for s on the inflow and outflow faces OA and BC. Once again we recall that (i) the boundary condition for ψ does not change under a given total flow, and (ii) a Dirichlet problem will always be obtained for ψ under boundary conditions of the second kind for s, which is the case in Problem 1.

Suppose only oil is recovered from the face BC. Then

$$k_1 p' \frac{\partial s}{\partial x} = -Q_0 \tag{3.1}$$

and hence, since $p' < 0$,

$$\frac{\partial s}{\partial x} > 0, \tag{3.2}$$

the sign of the derivative remaining fixed for all $t < t_0$. This means that the water saturation must be greater on the outflow face than at neighboring interior points. Therefore an increase in the water saturation at interior points due to an injection of water through the inflow face AO must lead to a still greater increase in s on BC, as follows from (3.2). We do not consider here the case when the saturation of the displacing phase varies on the displacement front from s_0 to some $s < \bar{s}$. Such a situation is a characteristic feature of the Buckley-Leverett flow model (see [5] and [6]), in which the capillary forces can be neglected. This model is normally realized in practice under large injection rates. Further, inasmuch as the relative permeability to the displaced phase tends to zero ($k_1 \rightarrow 0$) as saturation is approached, i.e. as $s \rightarrow \bar{s}$, it follows from (3.1) that the derivative $\partial s/\partial x$ must tend to infinity to guarantee a given choice of Q_0. This phenomenon is known as the end effect.

In the numerical solution of porous flow problems with boundary condition (3.1) the presence of the end effect makes it necessary to introduce a very fine mesh with respect to the space variables. For suppose the boundary condition (3.1) is approximated in the following way:

$$s_{Nj} = s_{N-1,j} - h \frac{Q_0}{k_1 p'},$$

where N is the number of mesh points with respect to the space variable x. The derivative $p' < 0$, and hence, if no restrictions need be imposed on the mesh width h with respect to the space variable x, for any water saturation approximation $s_{N-1,j}$ at an interior point, as saturation is approached, it is possible by virtue of the smallness of k_1 to find a value of h such that

$$-h \frac{Q_0}{k_1 p'} > \bar{s},$$

which is physically impossible. We will see below how important it is to restrict the magnitude of h.

Suppose both oil and water with a common total flow rate density Q_0 are recovered from the outflow face BC, the recovery of each phase occurring in proportion to its mobility. Then, as we saw in §2, $\partial s/\partial x = 0$. This boundary condition is also not completely obvious, at least after breakthrough of water at BC, since, as the water front approaches a production well, we will have $\partial s/\partial x < 0$ in a sufficiently small neighborhood of it. Thus more precise boundary conditions are needed at the production wells.

Conditions specifying that each phase is recovered in proportion to its flow rate (proposed by N. N. Janenko) can serve as one such refinement. It is clear in this connection that the total flow rate remains fixed.

At an injection well it is natural to specify the rate at which water is injected, which reduces to the condition

$$k_2 p' \frac{\partial s}{\partial x} = Q_0. \tag{3.3}$$

In the numerical realization of this boundary condition there arise the same difficulties as in the realization of boundary condition (3.1) on the outflow face. Inasmuch as the displacement process is initiated for values of s close to $\underline{s}$, we must reckon with the smallness of k_2. The simplest difference approximation of the boundary condition (3.3) gives

$$s_{0j} = s_{1j} - h \frac{Q_0}{k_2 p'}.$$

In order to ensure the fulfillment of the natural physical condition $s_{0,j} \leqslant \bar{s}$, we must choose h so that

$$h \frac{Q_0}{k_2 |p'|} \leqslant \bar{s}.$$

The latter condition is quite burdensome since, in order to satisfy it, h^*, i.e. the value of h near the boundary, must satisfy the condition $h^* \ll h$, where h is the mesh width with respect to the space variable x at interior points. Consequently, in refining the mesh near a boundary we introduce a parameter $\delta = h^*/h$. It turns out in this connection that the numerical solution of a porous flow problem with sources and sinks essentially depends on the parameter δ, which is defined near these sources and sinks. The result of a calculation of Problem 1 for different δ is shown in Figure 5.

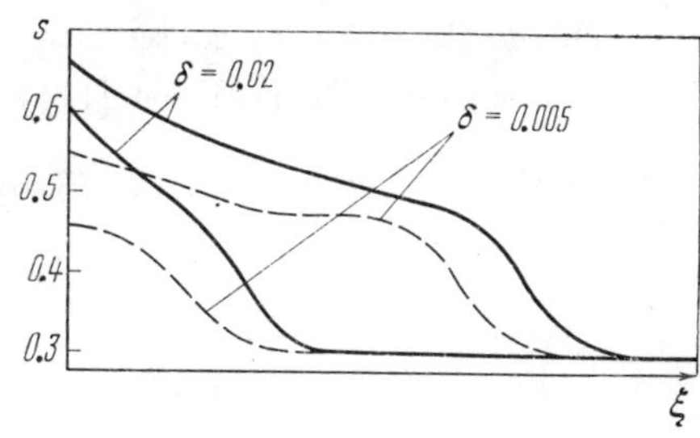

FIGURE 5

The boundary condition (2.15) was taken on the inflow face, while the boundary condition (2.22) was taken on the outflow face. The upper and lower boundaries of the stratum were assumed to be impermeable. The distribution of saturation s at a fixed moment of time t is given for $\eta = y/H = 0.5$ as a function of the dimensionless length $\xi = x/L$. Other methods of calculation for a boundary condition of type (3.3), e.g. the flux version of the double sweep method [7], possess an analogous singularity. The fact is that in all of these methods s is determined on the boundary, i.e. at the source or sink itself. We are then compelled to reckon with the fact that $\partial s/\partial x$ can be arbitrarily large and yet the calculated s must lie within the limits $\underline{s} \leqslant s \leqslant \bar{s}$.

We note that the vanishing of the relative permeabilities as s tends to its limiting values $\underline{s}$ and $\bar{s}$ does not lead to essential singularities in the numerical solution of equation (2.6) for the water saturation itself, since we obtain in this connection either the case $\partial s/\partial t = 0$ ($k_2 = 0$) or an ordinary differential equation for s ($k_1 = 0$). We also note that the dependence on δ of numerical solutions of porous flow problems with sources and sinks even in the absence of capillary forces was pointed out in [8]. Practical recommendations on the choice of h^* are given in [8] on the basis of a calculation of model problems having an analytic solution and a subsequent comparison of the numerical solution with the exact solution. The dependence on δ of the numerical solution of the P, R statement of a one-dimensional porous flow problem is removed in [9] by carrying out a special partitioning of the space mesh interval closest to the boundary.

We describe below a method of calculating s from equation (2.6) with a boundary condition of type (3.3) in which no restrictions on h^* arise. Inasmuch as (2.6) is numerically solved by applying an alternating direction method in which the solution of a two-dimensional problem is reduced to the successive solution of one-dimensional problems, we present the proposed method of calculating s by considering as an example the one-dimensional model equation

$$m\frac{\partial s}{\partial t} = -\frac{\partial}{\partial x}\left(\frac{k_1 k_2}{k_1 + k_2} p' \frac{\partial s}{\partial x}\right). \tag{3.4}$$

We pose a mixed Cauchy problem for equation (3.4), i.e. we specify s for $t = 0$ and the boundary conditions

$$k_2 p' \frac{\partial s}{\partial x}\bigg|_{x=0} = Q_0, \qquad \frac{\partial s}{\partial x}\bigg|_{x=1} = 0. \tag{3.5}$$

For a numerical solution of problem (3.4), (3.5) we choose the following conservative difference scheme with iterations with respect to the nonlinearity (see [10] and [11]):

$$A_i^{\nu\,n+1} s_{i+1}^{\nu+1\,n+1} - B_i^{\nu\,n+1} s_i^{\nu+1\,n+1} + C_i^{\nu\,n+1} s_{i-1}^{\nu+1\,n+1} + D_i^n = 0. \tag{3.6}$$

Here ν is the iteration index and

$$A_i^{\nu\,n+1} = -\frac{\tau}{h^2}\left(\frac{k_1 k_2}{k_1 + k_2} p'\right)_{i+1/2}^{\nu\,n+1},$$

$$C_i^{\nu\,n+1} = A_{i-1}^{\nu\,n+1}, \qquad B_i^{\nu\,n+1} = m + A_i^{\nu\,n+1} + C_i^{\nu\,n+1}, \qquad D_i^n = ms_i^n.$$

Inasmuch as $p' < 0$, it follows that $A_i > 0$, $C_i > 0$ and $B_i > A_i + C_i$, and system (3.6) with boundary conditions of the type

$$s_0 = \varkappa_1 s_1 + \omega_1, \qquad s_N = \varkappa_2 s_{N-1} + \omega_2$$

is solvable. Its solution can be found by means of the well-known double sweep method [12]:

$$X_i = \frac{A_i}{B_i - C_i X_{i-1}}, \qquad Y_i = \frac{D_i + C_i Y_{i-1}}{B_i - C_i X_{i-1}}, \qquad s_i = X_i s_{i+1} + Y_i.$$

To simplify the writing we omit some of the indices whenever it does not cause confusion. The usual approximation of the boundary condition for $x = 0$ in (3.5) is unsuitable for the determination of X_0 and Y_0 for reasons discussed above. We therefore proceed as follows. We write the difference equation for $i = 1$ (i.e. at the interior point closest to the inflow boundary):

$$m \frac{s_1^{n+1} - s_1^n}{\tau} = -\frac{1}{h}\left[\left(\frac{k_1}{k_1 + k_2} k_2 p' \frac{\partial s}{\partial x}\right)_{3/2}^{n+1} - \left(\frac{k_1}{k_1 + k_2} k_2 p' \frac{\partial s}{\partial x}\right)_{1/2}^{n+1}\right]. \tag{3.7}$$

The boundary condition at the inflow boundary will be considered to refer to the point $i = \frac{1}{2}$. Then in place of (3.7) we will have

$$m \frac{s_1^{n+1} - s_1^n}{\tau} = \frac{1}{h}\left(\frac{k_1}{k_1 + k_2}\right)_{1/2}^{n+1} Q_0 - \frac{1}{h}\left(\frac{k_1}{k_1 + k_2} k_2 p' \frac{\partial s}{\partial x}\right)_{3/2}^{n+1}. \tag{3.8}$$

The difference equation (3.8) connects s_1^{n+1} and s_2^{n+1}, which together with the equation

$$s_1^{n+1} = X_1 s_2^{n+1} + Y_1$$

permits one to determine X_1 and Y_1. We note that in using this method of calculation we do not determine s at the inflow boundary. This, however, does not prevent us from carrying out the calculation in the large, since we can refer all of the quantities calculated at the point $i = \frac{1}{2}$ to the point $i = 1$.

The algorithm presented above was perfected on both one-dimensional calculations for the model equation (3.4) and two-dimensional calculations in the s, ψ statement. These calculations confirmed the suitability of our algorithm for solving porous flow problems with a boundary condition of type (3.5) assigned at the inflow boundary. This makes it possible to solve the s, ψ and s, p_i statements of porous flow problems. We shall dwell below on some of the qualitative consequences of the calculations in the s, ψ statement.

But here we consider another method of describing a well that is suitable in the case when the well is modeled by a point source. Such a situation arises, for example, in an areal flooding problem involving a five-spot well pattern. The P, R statement of the two-dimensional case of this problem was considered in [1]. We present a method of calculating a well, which was proposed by the authors of [1], by considering as an example the one-dimensional model system

$$\frac{\partial}{\partial x}\left(M \frac{\partial P}{\partial x}\right) + \frac{\partial}{\partial x}\left(N \frac{\partial R}{\partial x}\right) = 0,$$

$$\frac{\partial}{\partial x}\left(N \frac{\partial P}{\partial x}\right) + \frac{\partial}{\partial x}\left(M \frac{\partial R}{\partial x}\right) = -4ms' \frac{\partial R}{\partial t}. \tag{3.9}$$

A solution of (3.9) for $t > 0$ is sought on the segment $a \leqslant x \leqslant b$, and at the boundary $x = a$ is located an injection well, while at the boundary $x = b$ is located a producing well of equal capacity. Suppose for the sake of definiteness that the injection well is fed only water, while both oil and water are recovered at the producing well in proportion to their mobilities. Boundary conditions at the wells can be formulated by assigning the flow rates of the corresponding phases. Then at the inflow boundary we shall have a boundary condition of type (3.3) with all of the attendant consequences.

In [1] the calculation for an oil reservoir is carried out as follows. At $x = a$ one uses the equations

$$\frac{\partial}{\partial x}\left(M\,\frac{\partial P}{\partial x}\right)+\frac{\partial}{\partial x}\left(N\,\frac{\partial R}{\partial x}\right)+\frac{G_2}{\Delta x}=0,$$

$$\frac{\partial}{\partial x}\left(N\,\frac{\partial P}{\partial x}\right)+\frac{\partial}{\partial x}\left(M\,\frac{\partial R}{\partial x}\right)-\frac{G_2}{\Delta x}=-4ms'\,\frac{\partial R}{\partial t}, \qquad (3.10)$$

with the additional conditions

$$P_{-1}=P_1, \qquad R_{-1}=R_1, \qquad (3.11)$$

which are valid in the presence of a point symmetry. At $x = b$ one uses the equations

$$\frac{\partial}{\partial x}\left(M\,\frac{\partial P}{\partial x}\right)+\frac{\partial}{\partial x}\left(N\,\frac{\partial R}{\partial x}\right)+\frac{G_1+G_2}{\Delta x}=0,$$

$$\frac{\partial}{\partial x}\left(N\,\frac{\partial P}{\partial x}\right)+\frac{\partial}{\partial x}\left(M\,\frac{\partial R}{\partial x}\right)-\frac{G_1-G_2}{\Delta x}=-4ms'\,\frac{\partial R}{\partial t} \qquad (3.12)$$

with the additional conditions

$$P_{N-1}=P_{N+1}, \qquad R_{N-1}=R_{N+1}. \qquad (3.13)$$

In (3.11) and (3.13) the lower index is the space index, while in (3.10) and (3.12) G_i is the volumetric flow rate per unit area of the ith phase and Δx is the space mesh increment. If we take (3.11) into account, the usual difference approximation of the second of equations (3.10) connects R_1 and R_0. This makes it possible to successively determine the double sweep coefficients X_i and Y_i. A difference approximation of the second of equations (3.12) together with (3.13) and the equation

$$R_{N-1}=X_{N-1}R_N+Y_{N-1}$$

makes it possible to determine R_N and then the remaining R_i.

The calculation of P proceeds analogously, although some singularities occur. The fact is that the system of algebraic equations obtained after a difference approximation of the first equations of (3.9), (3.10) and (3.12) with conditions (3.11) and (3.13) being taken into account does not have a unique solution, since we essentially have the one-dimensional analog of the Neumann problem for the determination of P. To obtain unique solvability of the problem one can fix P at the point $x = a$, i.e. assign a reference level for the total pressure of the two phases. The assignment of P_0 together with (3.11) and a difference approximation of the first of equations (3.10) permits us to determine P_1. The value of P_1 obtained in this way depends on $h = \Delta x$ and the given P_0, it being always possible to choose P_0 for a given h so that the sign of the derivative of P with respect to x does not change in a neighborhood of the injection well.

It remains to mention one more approach to the solution of porous flow problems with sources and sinks on the boundaries of a domain. This approach was first formulated in [13] in connection with the numerical solution of the problem of linear displacement of oil by water in which the influence of capillary forces are taken into account, and was later used in [14] to prove an existence theorem for the solution of a two-dimensional problem of porous flow of a two-phase incompressible fluid under certain additional restrictions. The essence of this approach consists in introducing in place of the water saturation a new unknown function which satisfies an equation of the same type as the equation satisfied by s but whose space derivative in a boundary condition of the second kind is bounded for any s. For example, one can introduce the function

$$r(s) = \int_{\underline{s}}^{s} k_1(\eta) k_2(\eta) \, | \, p'(\eta) \, | \, d\eta, \tag{3.14}$$

which satisfies a quasilinear parabolic equation of second order [14]. The equation for ψ need not be changed in this connection, since when $\underline{s} < s < \bar{s}$ we can use (3.14) to determine s from r. The boundary conditions (2.15) and (2.16) of Problem 1 are transformed into the following conditions for $r(s)$:

$$\left. \frac{\partial r}{\partial x} \right|_{OA} = k_2 Q_0, \qquad \left. \frac{\partial r}{\partial x} \right|_{BC} = -k_1 Q_0.$$

As s tends to its limiting values $\underline{s}$ and $\bar{s}$, the derivative $\partial r / \partial x$ tends to zero, which does not cause any complications in the numerical solution of the r, ψ statement of the problem.

More analogously to [13], one can also introduce the function

$$z(s) = \int_{\underline{s}}^{s} \frac{k_1(\eta) k_2(\eta)}{k_1(\eta) + k_2(\eta)} \, | \, p'(\eta) \, | \, d\eta. \tag{3.15}$$

which satisfies an equation of the same type as that satisfied by s. Here again the boundary conditions of the second kind for $z(s)$ corresponding to conditions (2.15) and (2.16) of Problem 1 do not have any singularities. But in the numerical solution of the r, ψ and z, ψ statements of porous flow problems we must reckon with the fact that the derivatives $\partial r / \partial s$ and $\partial z / \partial s$ tend to zero as s tends to its limiting values $\underline{s}$ and $\bar{s}$. Therefore the unique determination of the water saturation from (3.14) or (3.15) near its limiting values is a quite complicated problem.

§4. Some results of the numerical calculations

The numerical calculations in the s, ψ statement bore a systematic character and possessed the following basic purpose: to determine the behavior of a solution near the sources as well as the influence of the method of calculating a source on the behavior of a numerical solution in the large. A result of one of these calculations, which characterizes the dependence of a numerical solution on the parameter δ, was shown in Figure 5. In §3 we presented a method of calculating a source that permitted one to get rid of the restrictions on h^* in the calculation of the boundary condition (2.15) at the inflow boundary. This method was realized in a numerical solution of the s, ψ statement of Problem 1 with the boundary condition (2.22) at the outflow boundary. In this connection we made use of the functional dependencies of $f_1(s)$, $f_2(s)$ and $p(s)$ cited in [1] and the following input data: $\mu_1 = 0.17$ poises, $\mu_2 = 0.009$ poises, $k = 0.302 \times 10^{-7} \mathrm{cm}^2$, $L = 183 \, \mathrm{cm}$, $H = 15.24 \, \mathrm{cm}$, $Q_0 = 0.012 \, \mathrm{cm/sec}$, $m = 0.375$ and $b = 0.953 \, \mathrm{cm}$. Here $Q_0 = q_0 / bH$, where q_0 is the volumetric flow rate, H is the thickness of the stratum and b is the effective width of the stratum. The choice of the calculation parameters of the problem corresponds to a case close to pure imbibition, i.e. to a case in which the capillary forces essentially determine the structure of the flow.

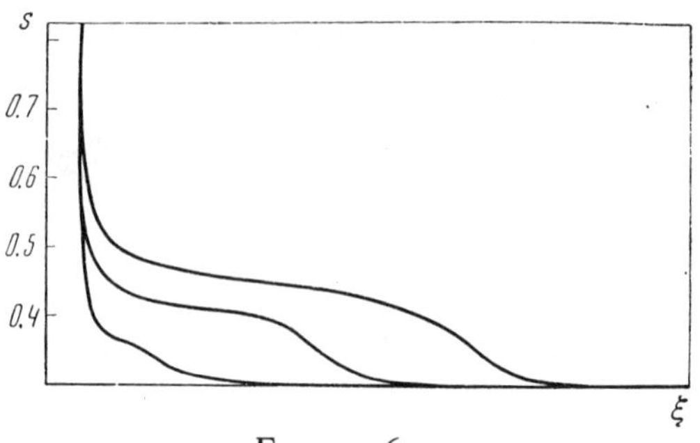

FIGURE 6

Before solving Problem 1, we reduced it to a dimensionless form, analogously to the way it was done in [13]. The numerical calculations showed the following. In the vicinity of the source (Figure 6) there is formed a domain of large gradients which lasts throughout the time of calculation, i.e. until the arrival of the water front at the face BC. The formation of a domain of large gradients in the vicinity of the source is by no means a calculational effect. For the calculation parameters used in Problem 1 the derivative $\partial s/\partial\xi$, $\xi = x/L$, takes the following values as a function of s:

s	0.2	0.3	0.4	0.5	0.6	0.7	0.8
$\partial s/\partial\xi$	-561.5	-227.3	-95.78	-66.29	-50.59	-27.92	-15.02

The mechanism of formation of the water front can be pictured by proceeding from the data of the calculation as follows. A watered zone is formed comparatively quickly in the vicinity of the well. The characteristic time of formation of the watered zone is denoted by t_1. We note that t_1 does not depend on L. The subsequent progress of the water front is basically due to imbibition, i.e. to the action of capillary forces. This explains the insignificant rise in water saturation behind the front when $t_1 \leqslant t \leqslant t_2$, where t_2 denotes the time of breakthrough at BC. We consider here the results of the calculation only for the first stage of the flow, when $t \leqslant t_2$, since it is only then that the effect of the influence of the source on the structure of the flow is revealed in a pure form. When $t > t_2$ the rise in water saturation in the stratum is affected by the boundary condition on BC. The time t_2 clearly depends on L, and for a real problem

$$\beta = \frac{t_2}{t_1} \gg 1. \tag{4.1}$$

For Problem 1 the parameter $\beta \simeq 7$, and the water saturation $s(t)$ at the boundary of

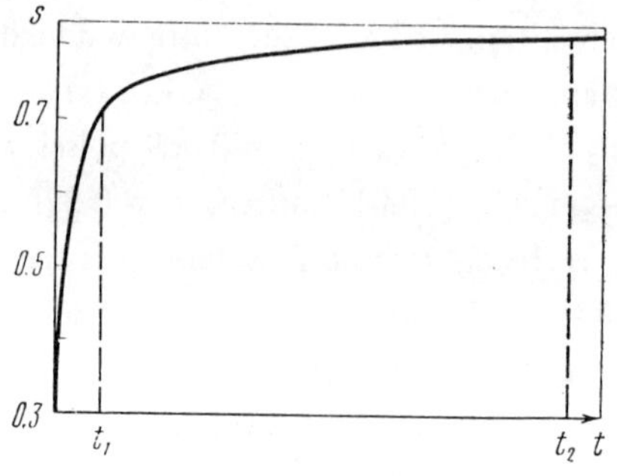

FIGURE 7

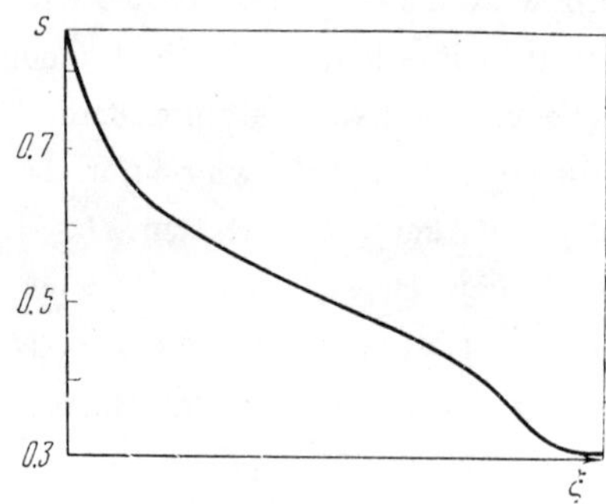

FIGURE 8

the watered zone is shown in Figure 7 for $\eta = y/H = 0.5$.

Thus the time mesh width τ in the numerical calculation is determined by the characteristic time t_1, and it follows from (4.1) that the time of calculation becomes inordinately large for a real problem. It would appear that the following way out is possible: to conduct the calculation with τ_1 up to the moment $t = t_1$ and then to increase τ to τ_2 so that $\tau_2/\tau_1 \simeq \beta$. But the fact is that, as the calculation shows, the domain of large gradients at the boundary of the watered zone lasts practically throughout the flooding process. Equation (2.6) is nonlinear, and hence at each time layer it is necessary to introduce iterations with respect to the nonlinearity that are similar to the method of successive approximations [10]. A rough criterion of convergence of the iterative process at a time layer can be written down as follows:

$$C\tau \,|\, \text{grad}\, s \,| < 1, \tag{4.2}$$

where C is a certain constant. From (4.2) and Figure 6 it follows that, other things being equal, the time mesh width τ is determined by the domain of large gradients. The introduction of local iterations with respect to the nonlinearity, which is so effective in one-dimensional problems, is impossible here because of logistic complexities.

An examination was made of the extent to which the boundary condition (2.15) is responsible for the domain of large gradients and its many consequences. Problem 1 was calculated with the boundary condition $s = \bar{s} = 0.85$ on the inflow face and the boundary condition on the outflow face unchanged. The distribution of saturation at breakthrough ($t = t_2$) is shown for $\eta = 0$ in Figure 8. Inasmuch as the new boundary condition coincides with the boundary condition for imbibition, it is not surprising that the result shown in Figure 8 is qualitatively in good agreement with the calculation in [15], where the case of pure imbibition was considered. We note that the calculation time for the whole problem with the new boundary condition was decreased by a factor of 10, due to an increase in τ.

In light of the above remarks it seems expedient to conduct the calculation of the watered zone separately, i.e. to divide the domain of porous flow into two domains. In one of the domains, which is free of sources, we will solve equations (2.6) and (2.7). In the domain containing the sources we will go over to another flow model. Such a division into two domains can be justified on physical grounds since the mechanism of formation of the watered zone and the mechanism of formation of the displacement front are essentially different. The matching of the solutions in these domains can be achieved by means of boundary conditions, it being highly desirable—and in this lies the reason for introducing two domains—to obtain in the domain free of sources a problem for s with a boundary condition of the first kind on the common boundary of the two domains.

We further show how one might pose Problem 1 with the introduction of two domains. As the flow model in the vicinity of a source, let us dwell for the sake of definiteness on a flow model for miscible fluids. We assume that the domain of flow is bounded by two coaxial cylinders and, to a first approximation, we will neglect the influence of the force of gravity in the watered zone. The mathematical formulation of the chosen flow model can be written down as follows (see [16]–[18]):

$$\frac{\partial s}{\partial t} + \frac{q_0}{2\pi r H m}\frac{\partial s}{\partial r} = \frac{q_0 \lambda}{2\pi r H m}\frac{\partial^2 s}{\partial r^2}. \tag{4.3}$$

Here r is the distance from the source and $\lambda = D/v$, where D is the molecular diffusivity and v is the rate of porous flow per unit area, which is given in the form

$$v = \frac{q_0}{2\pi m H r}. \tag{4.4}$$

Equation (4.3) will be solved in the domain $r_0 < r < R$ for a given initial distribution of saturation and the boundary condition

$$s = 1 \qquad \text{at} \qquad r = r_0. \tag{4.5}$$

The boundary condition for s at $r = R$ is obtained from the following matching condition for the zones of miscible and immiscible flows, in which we assume that the separation of the mixture into phases occurs under the action of capillary forces:

$$v_1 = -k_1 \frac{\partial p_1}{\partial r} = v\Big|_{r=R}, \qquad v_2 = -k_2 \frac{\partial p_2}{\partial r} = v\Big|_{r=R}.$$

On the other hand, for all $r \geqslant R$

$$\frac{\partial p_1}{\partial r} - \frac{\partial p_2}{\partial r} = p' \frac{\partial s}{\partial r}.$$

Therefore the second boundary condition for s takes the form

$$\frac{\partial s}{\partial r}\Big|_{r=R} = \frac{1}{p'}\left(\frac{1}{k_2} - \frac{1}{k_1}\right) v\Big|_{r=R}. \tag{4.6}$$

We note that boundary condition (2.15), which is inherently responsible for the domain of large gradients in Problem 1, can be obtained from (4.6) under the additional assumption that $k_2 \ll k_1$.

We also note that boundary condition (4.6) can be posed directly for Problem 1. Irrespective of the chosen model of flow in the watered zone, this condition means that the phase rates at the boundary are equal.

The matching of the solutions in the two domains can now be effected in the following manner. Suppose the water saturation at $r = R$ is equal to its initial value for $t < t^*$. Then problem (4.3), (4.5), (4.6) can be solved independently for $t < t^*$. Consider now the time $t \geqslant t^*$. Suppose s is found at $r = R$ from the solution of problem (4.3), (4.5), (4.6). With this value of s, we now solve equations (2.6) and (2.7). The function

$$\frac{1}{p'}\left(\frac{1}{k_2} - \frac{1}{k_1}\right)$$

of s in boundary condition (4.6) will be calculated for $t \geqslant t^*$ by means of the formula $s = \frac{1}{2}(s_{R+h_1} + s_{R-h_2})$, in which the indices 1 and 2 refer to the corresponding domains. It is clear that the time mesh width in each domain can be chosen independently, since the reduction of the boundary condition to a single moment of time can be achieved through interpolation.

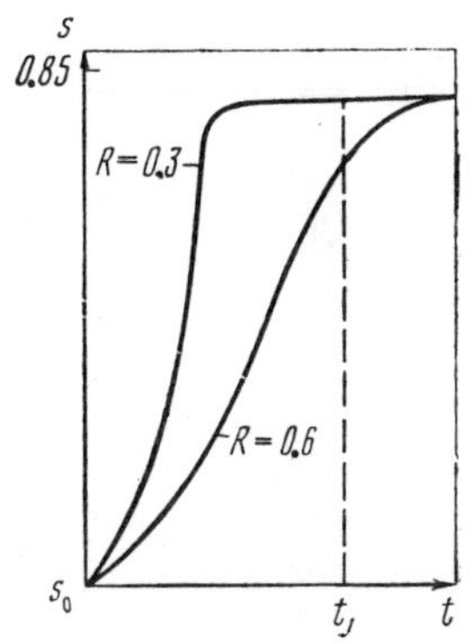

FIGURE 9

The calculations of the flow model for miscible fluids on the basis of equation (4.3) showed that the distribution of saturation with respect to length depends only insignificantly on the parameter λ (in the calculations λ varied from 0.05 cm to 1 cm). The water saturation $s(t)$ at $r = R$ basically depends on R; it is given for different R in Figure 9. It is very interesting that the time t in which s varies from s_0 to its limiting value $\bar{s} = 0.85$ turns out to be of the same order as t_1.

Another method of getting rid of the restrictions on τ due to the presence of the watered zone consists, apparently, in the possibility of making use of the fact that the time t_1 in a real problem is small in comparison with the time of the flooding process in the large, just as the width of the watered zone is small in comparison with the length of the stratum. But it is first necessary to determine how a variation of t_1 at the boundary influences the character of the flooding in the large. Such calculations were conducted on the basis of the one-dimensional Leas-Rapoport equation

$$m \frac{\partial s}{\partial t} + q(t) \frac{\partial}{\partial r} \frac{1}{r} \left(\frac{k_2}{k_1 + k_2} \right) + \frac{1}{r} \frac{\partial}{\partial r} \left(\frac{k_1 k_2}{k_1 + k_2} p' r \frac{\partial s}{\partial r} \right) = 0, \qquad (4.7)$$

which is written for the case of axial symmetry.

As a preliminary, equation (4.7) was reduced to dimensionless form and a mixed Cauchy problem with boundary conditions

$$s \big|_{\bar{r}=0,01} = s(t), \qquad \frac{\partial s}{\partial \bar{r}} \Big|_{\bar{r}=1} = 0 \qquad (4.8)$$

was posed for it. The functional dependencies for (4.7) and (4.8) were taken in the form

$$f_1(s) = \left(\frac{\bar{s} - s}{\bar{s} - \underline{s}} \right)^3, \qquad f_2(s) = \left(\frac{s - \underline{s}}{\bar{s}} \right)^3,$$

$$J(s) = 0.0072 \left(\frac{1}{s} - \frac{1}{\bar{s}} \right) + 0.5 \, (\bar{s} - s).$$

The value of $\bar{s}$ in these calculations was taken equal to 0.8.

In Figure 10 we show the various profiles of $s(t)$ at the injection well that were used in the calculation of problem (4.7), (4.8). In Figure 11 we show the profiles of $s(t)$ with different t_1 at various distances from the injection well. These preliminary calculations show that differences in $s(t)$ at the boundary apparently have a weak influence on the flooding process in the large. If we now managed to reduce the problem with boundary condition (2.15) to the problem with boundary condition $s = s(t)$ at the injection well, the restrictions on τ would apparently not arise, since the numerical methods being used are based on implicit difference schemes. Such a reduction is possible in principle if, for example, in equation (2.6) one introduces in the vicinity of the source fictitious sources with volumetric flow rate q_0 ensuring a given injection rate and at the same time assigns $s(t)$ at the well. This technique, as has already been mentioned, was used in [1], except, instead of assigning s at the source, symmetry conditions were employed (see (3.10)–(3.13)).

We mention in addition the calculations carried out to compare the various methods of linearization at the upper time layer of the implicit difference analog of the

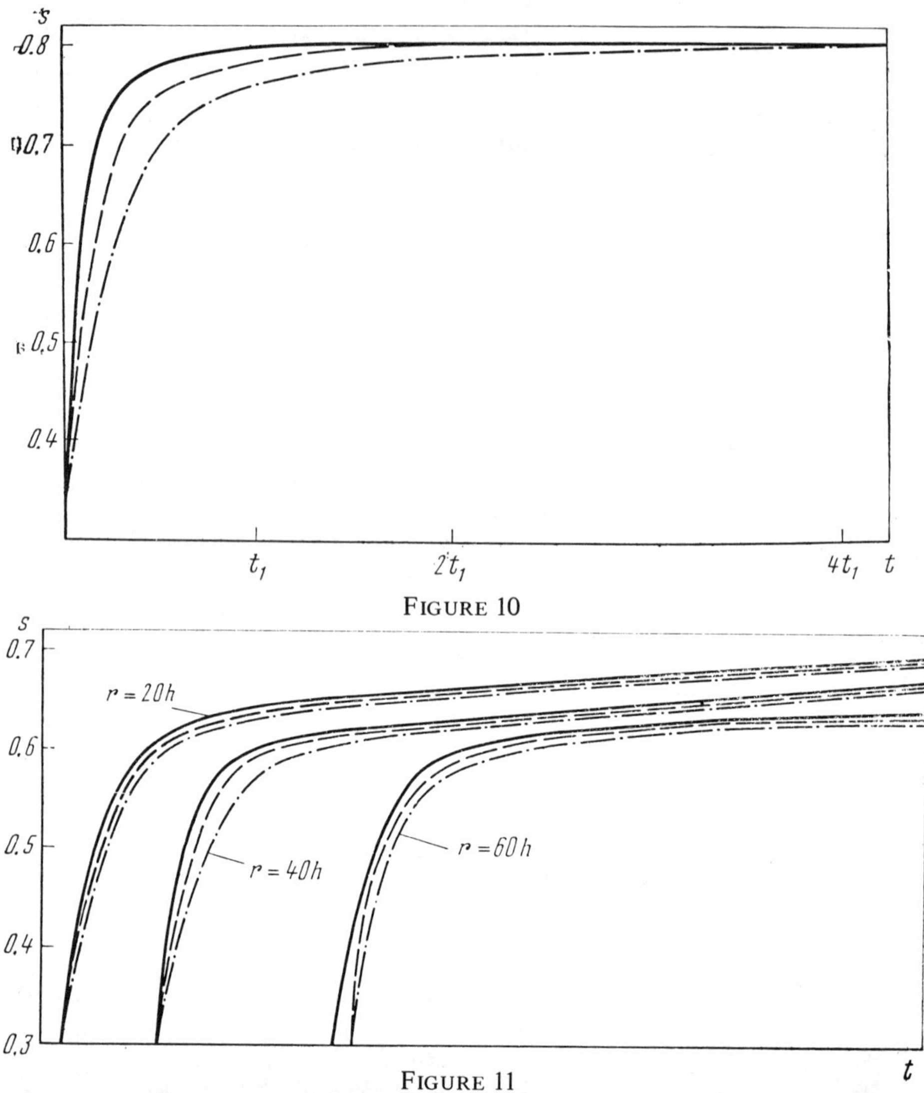

FIGURE 10

FIGURE 11

Leas-Rapoport equation (4.7). The linearization was conducted in two ways: by a simple
iteration method [10] and by Newton's method [19]. The calculations showed that
Newton's method converges more rapidly (two iterations at a time step instead of three to
four in the simple iteration method) but is more sensitive to the initial approximation, so
that a larger time step in comparison with the simple iteration method cannot be used.

    In conclusion, we cite the result of one more calculation illustrating the possibility of
a program realizing the s, ψ statement. The following problem was solved. The faces AB,
BB', OC and OA' (see Figure 2) are impermeable. Sources and sinks of equal strength are
uniformly distributed on $A'A$ and $B'C$. Water is injected through $A'A$. Both oil and water
are recovered through $B'C$ in proportion to their mobilities. The input calculation parameters are the same as in Problem 1. The lines of flow of this problem at a particular
moment of time are shown in Figure 12.

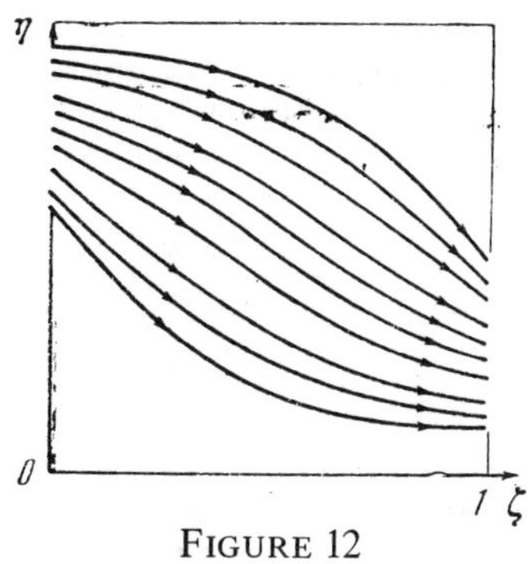

FIGURE 12

The calcualtion of system (1.13) was done by N. M. Salimova, the calculation of equation (4.7), by V. V. Salovatkin, and the calculations to compare the linearization methods, by V. M. Kutnjašenko. All of the remaining calculations were done by È. V. Smirnova. An active role in the discussion of the results of this work was taken by V. L. Danilov, B. G. Kuznecov and N. N. Janenko. The author is deeply grateful to them for valuable suggestions and remarks.

BIBLIOGRAPHY

1. Jim Douglas, Jr., D. W. Peaceman and H. H. Rachford, Jr., *A method for calculating multi-dimensional immiscible displacement*, AIME Tech. Paper 8090; abstract, J. Petroleum Technology, October, 1959, 85–86.

2. A. N. Konovalov, *On the question of the numerical solution of multidimensional problems on filtration of a two-phase incompressible fluid*, Izv. Sibirsk. Otdel. Akad. Nauk SSSR **1970**, no. 8, 46–54. (Russian)   MR **46** #505.

3. V. L. Danilov, A. N. Konovalov and S. A. Jakuba, *Equations and boundary-value problems in the theory of two-phase filtration flow in a porous medium*, Dokl. Akad. Nauk SSSR **183** (1968), 307–310 = Soviet Phys. Dokl. **13** (1968/69), 1102–1104.

4. N. N. Janenko, *The method of fractional steps for solving multidimensional problems of mathematical physics*, "Nauka", Novosibirsk, 1967; English transl., Springer-Verlag, Berlin and New York, 1971.   MR **36** #4815; **46** #6613.

5. R. E. Collins, *Flow of fluids through porous materials*, Reinhold, New York, 1961.

6. I. A. Čarnyĭ, *Underground hydrogasdynamics*, Gostoptehizdat, Moscow, 1963. (Russian)

7. L. M. Degtjarev and A. P. Favorskiĭ, *A flux version of the double sweep method*, Ž. Vyčisl. Mat. i Mat. Fiz. **8** (1968), 679–684 = USSR Comput. Math. and Math. Phys. **8** (1968), no. 3, 252–261.   MR **38** #4058.

8. G. G. Vahitov, *Effective methods for the solution of problems of the development of inhomogeneous oil-bearing strata by the method of finite differences*, Gostoptehizdat, Moscow, 1963. (Russian)

9. M. A. Tairov, *Solution of a nonlinear problem of two-phase flow in a porous medium*, Ž. Vyčisl. Mat. i Mat. Fiz. **6** (1966), 106–112 = USSR Comput. Math. and Math. Phys. **6** (1966), no. 1, 150–159.

10. A. A. Samarskiĭ and I. M. Sobol', *Examples of numerical calculation of temperature waves*, Ž. Vyčisl. Mat. i Mat. Fiz. **3** (1963), 702–719 = USSR Comput. Math. and Math. Phys. **3** (1963), 945–970.   MR **27** #3089.

11. A. A. Samarskiĭ, *Lectures on the theory of difference schemes*, Moscow, 1969. (Russian)

12. I. S. Berezin and N. P. Židkov, *Computing methods*. Vol. 2, Fizmatgiz, Moscow, 1959; English transl., Pergamon Press, Oxford; Addison-Wesley, Reading, Mass., 1965.   MR **22** #12686; **31** #1756.

13. Jim Douglas, Jr., P. M. Blair and R. J. Wagner, *Calculation of linear waterflood behavior including the effects of capillary pressure*, AIME Tech. Paper 8013; abstract, J. Petroleum Technology, June, 1958, 131a.

14. S. N. Antoncev and V. N. Monahov, *On some problems of porous flow of two-phase incompressible fluids*, Dinamika Splošnoĭ Sredy, vyp. 2, "Nauka", Novosibirsk, 1969, pp. 156–167. (Russian)

15. P. M. Blair, *Calculation of oil displacement by intercurrent water imbibition*, Paper no. 1475G, AIME Fourth Biennial Secondary Recovery Sympos., Wichita Falls, Texas, May, 1960.

16. E. V. Tesljuk, G. F. Trebin and Ju. M. Ostrovskiĭ, *On the porous flow of a mixture of fluids under conditions of plane radial flow and tubular flow of variable cross-section*, Trudy Vsesojuz. Neftegaz. Naučno-Issled. Inst. vyp. **40** (1963), 115–137. (Russian) RŽMeh. 1964 #3Б856.

17. E. V. Tesljuk, G. F. Trebin and Ju. M. Ostrovskiĭ, *Theoretical study of the porous flow of miscible fluids*, Trudy Vsesojuz. Neftegaz. Naučno-Issled. Inst. vyp. 42 (1965), 174–180. (Russian) RŽMeh. 1966 #2Б1066.

18. P. I. Zabrodin, N. L. Rakovskiĭ and M. D. Rozenberg, *Displacement of oil from strata by solvents*, "Nedra", Moscow, 1968. (Russian)

19. L. V. Kantorovič, *On Newton's method*, Trudy Mat. Inst. Stekov. 28 (1949), 104–144. (Russian)   MR 12, 419.

Trudy Mat. Inst. Steklov.
122 (1973)

Proc. Steklov Inst. Math.
122 (1973)

NONADIABATIC MOTIONS IN A PERFECT GAS
(AUTOMODEL SOLUTIONS)

UDC 518.519.951

V. E. NEUVAŽAEV

ABSTRACT. The present paper is devoted to a study of one-dimensional non-adiabatic motions of a gas that takes into account the influence of nonlinear heat conduction and energy release.

Bibliography: 17 items.

It is well known that it is extremely difficult to solve exactly the general problems mentioned in the preceding abstract under arbitrary initial and boundary conditions. But in a whole series of cases certain problems admit well-known similarity transformations [1], [2] permitting one to reduce the system of partial differential equations describing the motion to a system of ordinary differential equations. The study of self-similar, or automodel, motions is very important for the determination of the structure of the resulting solutions inasmuch as the solutions in question are, as a rule, discontinuous. In addition, the automodel solutions can serve as a criterion of accuracy for the difference schemes being applied.

The study of automodel solutions has received considerable attention (see, for example, the bibliography in the book [3]). In the problems considered the resulting ordinary differential equations are either integrated in closed form or reduced to the investigation of a single equation. The determination of the solution by a numerical procedure is not essentially difficult and can always be carried out by well-known methods.

The automodel motions studied in the present paper reduce to not one, but a system of nonlinear differential equations. For the latter we pose a boundary value problem with singular boundary conditions, the numerical solution of which involves considerable difficulties and cannot be realized by the usual methods. The difficulties are well known [4]: if the integral curves diverge rapidly in the direction of integration, the errors admitted under the calculation at each successive step increase sharply. In the boundary value problems in question, where conditions are given on two boundaries, the integral curves diverge not just in the direction of one boundary but also in the directions of the other. A study of the behavior of the integral curves, which involves a determination of the character of the singular points, permits one to construct a convergent iterative process that leads to the solution of the problem with a given accuracy.

AMS (MOS) subject classifications (1970). Primary 76G15.

§1 is of an auxiliary character. It contains a general statement of the problem and the mathematical apparatus that will be used to investigate it.

In §2 a solution of the problem of outflow of a gas into a vacuum under the action of a power law energy release is presented in Eulerian coordinates. A statement of the problem and the difficulties encountered in solving it were indicated by E. I. Zababahin. A brief presentation of the method in Lagrangian coordinates was published in [5].

§3 contains an automodel problem on the motion of a heat-conducting gas when the temperature is prescribed with a power law time dependence on the boundary with a vacuum. This case is a limiting case for the problem with "piston" discussed in [6] and reduces under numerical integration to the singularities mentioned above. The results of this section were presented in [7].

§4 contains the solution of a problem on an explosion in a heat-conducting gas under spherical symmetry. The problem was solved under the assumption of a discontinuity in the temperature and heat flow on the disturbance front by V. P. Korobeĭnikov in [8]. This solution was later repeated by L. Elliott [9]. In [10] the author showed that such solutions are unstable, and found a stable solution under the assumption that the temperature is continuous. Here, in conjunction with a more complete presentation of the problem, we adduce graphs of a comparison with a numerical calculation carried out with the use of the difference method of [11].

The author wishes to convey his deep gratitude to his scientific mentor N. N. Janenko. The author acknowledges his indebtedness to M. N. Nečaev, P. P. Volosevič and S. P. Kurdjumov for useful discussion, and thanks T. G. Ivčenko for his help in carrying out the calculations.

§1. General statement of the problem and the mathematical apparatus

1. *The basic equations in Eulerian and Lagrangian coordinates.* We will consider one-dimensional motions of a perfect gas with the equations of state

$$p = R\rho T, \qquad \varepsilon = \frac{R}{\gamma - 1} T, \qquad (1.1.1)$$

where R is the gas constant, γ is the ratio of specific heats, p is the pressure, ρ is the density, T is the absolute temperature and ϵ is the internal energy.

In Eulerian variables, the equations of motion expressing the three basic laws of mechanics are the momentum equation

$$\rho \left(\frac{\partial u}{\partial t} + u \frac{\partial u}{\partial r} \right) + \frac{\partial p}{\partial r} = 0, \qquad (1.1.2)$$

the continuity equation

$$\frac{\partial \rho}{\partial t} + \frac{\partial (\rho u)}{\partial r} + \frac{\nu \rho u}{r} = 0, \qquad (1.1.3)$$

and the energy equation

$$\frac{\partial}{\partial t} \left(\varepsilon + \frac{u^2}{2} \right) + u \frac{\partial}{\partial r} \left(\varepsilon + \frac{u^2}{2} \right) + \frac{\partial}{\rho r^\nu \partial r} r^\nu \left[pu - \varkappa \frac{\partial T}{\partial r} \right] = \frac{\partial F}{\partial t} + u \frac{\partial F}{\partial r}. \qquad (1.1.4)$$

Here r is the distance from the center of symmetry to a given point, t is the time, u is the velocity, κ is the coefficient of thermal conductivity, F is the energy release per unit mass, $\nu = 2$ for the case of spherical symmetry, $\nu = 1$ for the case of cylindrical symmetry and $\nu = 0$ for the case of plane symmetry.

Inasmuch as the subject of the present paper is a study of the nonadiabatic motions of a gas, the equation of conservation of energy contains terms characterizing the influx of heat due to heat conduction and energy release.

Using the mass variables of Lagrange, equations (1.1.2)–(1.1.4) can be transformed into the equations

$$\frac{du}{dt} + r^{\nu}\frac{\partial p}{\partial m} = 0, \quad \frac{d\,(1/\rho)}{dt} - \frac{\partial\,(r^{\nu}u)}{\partial m} = 0,$$

$$\frac{d}{dt}\left(\varepsilon + \frac{u^2}{2}\right) + \frac{\partial}{\partial m}r^{\nu}\left[pu - \kappa\rho r^{\nu}\frac{\partial T}{\partial m}\right] = \frac{dF}{dt}. \tag{1.1.5}$$

The systems of equations (1.1.2)–(1.1.4) and (1.1.5) are systems of quasilinear equations the integration of which generally encounters significant difficulties.

2. *Equations in self-similar variables.* In the sequel we will always assume that at $t = 0$ the gas is at rest, and has a certain initial density and zero temperature:

$$t = 0: \quad u = 0; \quad \rho = \rho_0, \quad T = 0. \tag{1.2.1}$$

In order for the problem to be self-similar it is necessary on the basis of dimensional analysis [1] that the coefficient of thermal conductivity κ and the energy release function F have power law dependencies; for example,

$$\kappa = \kappa_0 \frac{T^m}{\rho^l}, \tag{1.2.2}$$

$$F = F_0 t^n, \tag{1.2.3}$$

where κ_0 and F_0 are physical parameters and m, l and n are numerical parameters with $m > 0$ and $l > 0$. If we now put

$$m = \frac{1+n}{n}, \quad \lambda = \frac{r\,(n+2)}{2\,\sqrt{RB_0}\,t^{1+n/2}},$$

$$T = B_0 t^n \theta\,(\lambda), \quad u = \sqrt{RB_0}\,t^{n/2}\zeta\,(\lambda), \quad \rho = \rho_0 \delta\,(\lambda), \tag{1.2.4}$$

where B_0 is a certain physical parameter, the system of equations (1.1.2)–(1.1.4) can be transformed with the use of (1.2.4) into the system

$$\delta\left[\frac{n}{n+2}\zeta + (\zeta - \lambda)\,\zeta'\right] + (\delta\theta)' = 0, \tag{1.2.5}$$

$$-\lambda\delta' + (\delta\zeta)' + \frac{\nu\delta\zeta}{\lambda} = 0, \tag{1.2.6}$$

$$\frac{2n}{(n+2)\,(\gamma-1)} = \frac{2n}{n+2}\left(\frac{\zeta^2}{2} + \frac{\theta}{\gamma-1}\right) + (\zeta - \lambda)\left(\frac{\zeta^2}{2} + \frac{\theta}{\gamma-1}\right)'$$

$$+ \frac{1}{\lambda^{\nu}\delta}\left[\lambda^{\nu}\left(\delta\theta\zeta - \frac{n+2}{2}A\,\frac{\theta^{\frac{1+n}{n}}}{\delta^l}\,\theta'\right)\right]. \tag{1.2.7}$$

Here A is the numerical parameter

$$A = \frac{\varkappa_0 B_0^{1/n}}{R^2 \rho_0^{l+1}},$$ (1.2.8)

while the physical parameters B_0 and F_0 are connected by the relation

$$F_0 = \frac{R}{\gamma - 1} B_0.$$ (1.2.9)

The latter is implied by the following condition, which is valid for that part of the gas to which the disturbance has not yet propagated:

$$\varepsilon = \frac{RT}{\gamma - 1} = F_0 t^n.$$

3. *Boundary conditions.* With the use of (1.2.4), (1.2.8) and (1.2.9), we can establish what the boundary conditions must be in order for the problem to be an automodel problem.

Under the initial conditions (1.2.1) the most general boundary conditions will be the conditions on a "piston" if one assumes that the "piston" moves from the center of symmetry according to the power law

$$u = U_0 t^{n/2}$$ (1.3.1)

with a temperature

$$T = T_0 t^n$$ (1.3.2)

assigned on it.

It is easy to obtain the conditions on a "piston" in dimensionless variables. As a result of the fact that the "piston" moves along a trajectory that is at the same time a λ-line, we will have the λ-line condition

$$r = \frac{2}{n+2} \sqrt{RB_0}\, t^{1+n/2} \lambda_0,$$

and the trajectory condition

$$u = \frac{dr}{dt} = \sqrt{RB_0}\, t^{n/2} \lambda_0.$$

Taking (1.2.4) into account, we get

$$\lambda_0 = \zeta_0.$$ (1.3.3)

From (1.3.1) and (1.3.2) it follows that for $\lambda = \lambda_0$

$$\zeta_0 = \frac{U_0}{\sqrt{RB_0}}, \qquad \theta_0 = \frac{T_0}{B_0}.$$ (1.3.4)

Consequently the conditions on a "piston" correspond to relations (1.3.3) and (1.3.4), in which the numerical velocity ζ_0 and the numerical temperature θ_0 are known constants determined by the initial parameters of the problem. On the front of the disturbance (or disturbance front) we will always have a background, i.e.

$$u = 0, \qquad T = \frac{(\gamma - 1)\, F_0}{R}\, t^n, \qquad \rho = \rho_0, \qquad (1.3.5)$$

or, in numerical variables,

$$\lambda = \lambda_f: \qquad \zeta = 0, \qquad \theta = 1, \qquad \delta = 1. \qquad (1.3.6)$$

The index f means that the quantity is referred to the front.

In the sequel the conditions (1.3.3) and (1.3.4) on a "piston" are not used in their general forms; in fact, only the following special cases are considered: the plane case ($\nu = 0$) when the "piston" is a boundary with a vacuum:

$$\lambda_0 = \zeta_0, \qquad \theta_0 \delta = 0, \qquad (1.3.7)$$

and the cylindrical and spherical cases ($\nu = 1$ and $\nu = 2$) when the "piston" is the center of symmetry, i.e. is at rest:

$$\lambda_0 = \zeta_0 = 0. \qquad (1.3.8)$$

To the indicated boundary conditions, as will be seen below, correspond singular points of the system of ordinary equations (1.2.5)–(1.2.7), whereas, in the general case, the conditions on a "piston" are not singular. Here and below the boundary conditions will be called singular if, upon substituting them into the system of equations (1.2.5)–(1.2.7) solved for the derivatives, the numerators and denominators simultaneously vanish. The boundary conditions can compose a manifold of points in the appropriate space.

The disturbance front (1.3.6), which is not singular if one takes into account both heat conduction and energy release, will be singular if one only takes into account either heat conduction or energy release.

Thus the special boundary conditions (1.3.7) and (1.3.8) corresponding to outflow into a vacuum and a center of symmetry reduce to two-point boundary value problems in which the left and right boundaries are singular.

In the case of numerical solution of the system of ordinary equations (1.2.5)–(1.2.7) the above-mentioned singular conditions lead to difficulties associated with a loss of accuracy (see §1.5). A loss of accuracy also occurs under a numerical integration in the case of general conditions on a "piston" that are similar to conditions (1.3.7) or (1.3.8). This remark also applies to the conditions on a front. Therefore, by considering special cases, we can indicate a method of solution of the problem in the general case.

4. *Conditions on a discontinuity. Isothermal jump.* It is well known that a solution of the system (1.1.2)–(1.1.4) can be discontinuous. The basic laws of conservation of mass, momentum and energy must hold across a surface of strong discontinuity.

Letting D denote the speed of the discontinuity, the index 1 denote a quantity in front of the discontinuity and the index 2 a quantity behind the discontinuity, we write these conditions in the form

$$\rho_1(u_1 - D) = \rho_2(u_2 - D), \qquad \rho_1(u_1 - D)^2 + R\rho_1 T_1 = \rho_2(u_2 - D)^2 + R\rho_2 T_1,$$

$$\rho_1(u_1 - D)\left(\frac{u_1^2}{2} + \frac{RT_1}{\gamma - 1}\right) + R\rho_1 T_1 u_1 - \varkappa_1 \frac{\partial T}{\partial r}\bigg|_1$$

$$= \rho_2(u_2 - D)\left(\frac{u_2^2}{2} + \frac{RT_1}{\gamma - 1}\right) + R\rho_2 T_1 u_2 - \varkappa_2 \frac{\partial T}{\partial r}\bigg|_2. \qquad (1.\,4.\,1)$$

These equations are derived by assuming that the temperature is continuous across a discontinuity (isothermality) in the presence of heat conduction. If this assumption is not used, the problem turns out to be indeterminate. This will be considered in detail in §4.4.

Since we are studying automodel problems, a discontinuity will always be a λ-line and hence

$$D = \sqrt{RB_0}\, t^{n/2} \lambda_1. \qquad (1.\,4.\,2)$$

The substitution of (1.2.4) and (1.4.2) into (1.4.1) gives

$$\zeta_2 = \lambda_1 + \frac{\theta_1}{\zeta_1 - \lambda_1}, \qquad \delta_2 = \frac{\delta_1(\zeta_1 - \lambda_1)^2}{\theta_1},$$

$$q_2 = q_1 + \frac{\lambda_1^\gamma}{2 + n}\, \delta_1 \frac{(\zeta_1 - \lambda_1)^4 - \theta_1^2}{\zeta_1 - \lambda_1}, \qquad q = -\frac{A}{\delta^l} \lambda^\nu \theta^{\frac{1+n}{n}} \theta'. \qquad (1.\,4.\,3)$$

5. *Remarks on the method of solution.* The notion of stability of a solution plays an important role in the theory of ordinary differential equations.

In the case of the elementary linear equation

$$\frac{dx}{dt} = ax, \qquad x(0) = 0, \qquad t \geqslant 0, \qquad (1.\,5.\,1)$$

where a is a constant, it is known that the solution is stable for $a < 0$ and unstable for $a > 0$; in fact, to a perturbation $x(0) = \epsilon$ of the initial data there corresponds the solution

$$x = \varepsilon e^{at}, \qquad (1.\,5.\,2)$$

which tends to zero with increasing t for $a < 0$ and grows unboundedly with increasing t for $a > 0$.

If problem (1.5.1) is solved for $t \leqslant 0$, the solution will be stable for $a > 0$ and unstable for $a < 0$. This fact is connected with the behavior of the integral curves of the equation of (1.5.1). The solution is stable if the integral curves converge in the direction of integration, and unstable if they diverge in the direction of integration.

The direction of stability becomes particularly important in the numerical integration of an equation. A calculation in an unstable direction leads to an increase in the errors admitted in the calculation process. Therefore a numerical integration should be conducted in a stable direction. In the above example for a single equation, such a direction can always be found.

We note that if the curves diverge slowly in a given interval $[t_0,\, T]$, a numerical integration can also be conducted in an unstable direction; the loss in accuracy that will occur can be compensated for by increasing the accuracy of the calculation. But since, in practice, the accuracy cannot be unlimitedly increased, in the case of rapidly diverging curves such a method often does not ensure the desired number of significant figures and, as a rule, turns out to be unsatisfactory.

The problem of choosing a stable direction is significantly more complicated in the case of a Cauchy problem or boundary value problem posed for a system of equations. For example, in the case of a system of elementary linear equations that reduces to the single second-order equation

$$\frac{d^2x}{dt^2} - k^2x = f, \tag{1.5.3}$$

where k^2 and f are constants, an error $\delta x = \epsilon$ satisfies the equation

$$\frac{d^2\varepsilon}{dt^2} - k^2\varepsilon = 0,$$

the general integral of which is

$$\varepsilon = C_1 e^{kt} + C_2 e^{-kt},$$

i.e. an error in the solution increases in any direction. Such an equation can be solved by the well-known "double sweep" method [12], the idea of which consists in reducing (1.5.3) to a system of first order equations for each of which a stable direction of integration can be chosen.

The system of ordinary equations (1.2.5)–(1.2.7) under consideration is essentially nonlinear. A boundary value problem with conditions on a disturbance front and on a "piston" is posed for it. The problem cannot be solved by means of quadratures, and is integrated numerically. We construct below an iterative method for solving boundary value problems, the ideal of which consists in choosing variables which permit, as in the "double sweep" method, each equation of the system to be integrated in a stable direction.

Let us explain it by an example. Suppose given a boundary value problem for the system of two nonlinear equations

$$\frac{dy_1}{dt} = Y_1(t,\ y_1,\ y_2), \tag{1.5.4}$$

$$\frac{dy_2}{dt} = Y_2(t,\ y_1,\ y_2) \tag{1.5.5}$$

in which the boundary conditions are

$$\Phi[y_1(0),\ y_2(0)] = 0, \tag{1.5.6}$$

$$\Psi[y_1(T),\ y_2(T)] = 0. \tag{1.5.7}$$

We assume that system (1.5.4), (1.5.5) is such that roundoff errors rapidly increase under an integration of the system both from $t = 0$ to $t = T$ and from $t = T$ to $t = 0$. It is obvious that this problem cannot be solved by the shooting method. One therefore seeks new variables (it will be assumed that the original variables t, y_1, y_2 are already the new variables) such that directions of stable integration can be found for equation (1.5.4) when $y_2(t)$ is preassigned, and for (1.5.5) when $y_1(t)$ is preassigned.

In the $(t,\ y_1,\ y_2)$ space the assignment of $y_2(t)$ means the projection of the integral curves of equation (1.5.4) onto the surface $y_2 = y_2(t)$. Analogously, the assignment of $y_1(t)$ for (1.5.5) corresponds to the projection of the integral curves of (1.5.5) onto the corresponding surface $y_1 = y_1(t)$.

Thus it is assumed that for each projected family there exists a stable direction that remains stable over all of a given interval $[0, T]$.

Suppose for the sake of definiteness that (1.5.4) is stably integrated from $t = 0$ to $t = T$ while (1.5.5) is stably integrated from $t = T$ to $t = 0$. Then the following iterative process can be constructed for system (1.5.4), (1.5.5).

For the first iteration over the interval $[0, T]$ we assign, for example, the function $y_2^0(t)$. We determine $y_1^0(0)$ from condition (1.5.6) and find $y_1^0(t)$ by numerically integrating (1.5.4). We determine $y_2^1(T)$ from condition (1.5.7) and $y_1^0(T)$, and find the next approximation $y_2^1(t)$ of $y_2(t)$ by integrating (1.5.5). The process is repeated for the new function $y_2^1(t)$, and so on until convergence. If the chosen directions of integration are stable, the process will converge.

The difficulty in applying the above method obviously lies in the determination of the new variables. In the problems investigated below the new variables are found by studying the behavior of the integral curves. The boundary conditions are singular, and therefore one usually determines the behavior of the curves in a domain that permits one to choose the desired variables.

6. *Method of investigating singular points.* The singular points under investigation of the nonlinear system (1.2.5)–(1.2.7) are nonisolated singular points. In this case the corresponding linearized system does not always completely determine the qualitative behavior of the integral curves [13]. But we will be interested in the behavior of the curves in a specific domain not containing the singular direction associated with the manifold of singular points. In such cases an investigation of the linearized equations is sometimes sufficient (§2.2). If, on the other hand, the linear terms do not yield anything, one takes into account the nonlinear terms, solving the system for the derivatives and retaining the principal terms in the right sides. One then investigates the resultant system.

For a single equation it has been proved [14] that the equation usually obtained in this way yields a complete qualitative picture of the behavior of the integral curves. The other cases encountered by us will also be considered by following [14].

Suppose given the equation

$$y' = \frac{P(x, y)}{Q(x, y)}. \tag{1.6.1}$$

The point $x = 0$, $y = 0$ is a singular point, i.e. $P(0, 0) = Q(0, 0) = 0$. Setting $y/x = u$, we form the function

$$\psi(u, x) = \frac{P(x, ux)}{Q(x, ux)} - u$$

and, solving the equation

$$\psi(u, 0) = 0, \tag{1.6.2}$$

we find the directions along which the integral curves can reach the singular point. Such directions are called critical.

Consider the case when there exist a finite number of real critical directions, resulting in the case of simple roots of (1.6.2). Suppose the critical direction being investigated is $y/x = u_1$. We will assume that the direction u_1 does not satisfy the equation $Q(x, y) = 0$. Such a direction is called an ordinary simple critical direction. In this case the sign of the derivative $\partial\psi(u_1, 0)/\partial u$ determines the behavior of the integral curves in a sector containing the critical direction. If $\partial\psi(u_1, 0)/\partial u > 0$, an uncountable number of integral curves enter the singular point with slope u_1. If $\partial\psi(u_1, 0)/\partial u < 0$, the slope u_1 has one and only one integral curve.

There is no general theory on the basis of which one can be certain that the qualitative behavior of the curves of a system of ordinary equations all of whose characteristic roots are equal to zero will coincide with the behavior of the curves of the truncated system obtained by retaining the principal terms in the numerator and denominator. But by analogy with a single equation there is reason to hope that a family of curves entering the singular point will belong to the original system if it belongs to the truncated system. Such an assertion always holds for a single equation.

§2. Outflow of gas into a vacuum under the action of energy release

1. *Equations and boundary conditions.* In §1 we obtained the general equations in automodel variables, (1.2.5)–(1.2.7), for a heat-conducting and energy-releasing gas. Consider the case when heat conduction is not taken into account. Then for the plane case equations (1.2.5)–(1.2.7) take the form

$$\frac{n}{n+2}\zeta + (\zeta - \lambda)\zeta' + \theta\frac{\delta'}{\delta} + \theta' = 0, \tag{2.1.1}$$

$$(\lambda - \zeta)\delta' = \zeta'\delta, \tag{2.1.2}$$

$$\frac{2n}{(n+2)(\gamma-1)} = \frac{2n}{n+2}\left(\frac{\zeta^2}{2} + \frac{\theta}{\gamma-1}\right) + (\zeta - \lambda)\left(\frac{\zeta^2}{2} + \frac{\theta}{\gamma-1}\right)' + \theta\zeta\frac{\delta'}{\delta} + (\theta\zeta)'. \tag{2.1.3}$$

It is easily seen that one can use (2.1.2) to eliminate the expression δ'/δ from equations (2.1.1) and (2.1.3) and obtain a system of two equations for the unknowns ζ and θ that does not depend on δ. When solved for the derivatives, these equations take the form

$$\zeta' = \frac{N[2(\theta - 1) + \zeta(\lambda - \zeta)]}{(\lambda - \zeta)^2 - \gamma\theta} = \frac{(\theta' + N\zeta)(\lambda - \zeta)}{(\zeta - \lambda)^2 - \theta}, \tag{2.1.4}$$

$$\theta' = \frac{N}{\lambda - \zeta}\frac{2(1 - \theta)[\theta - (\lambda - \zeta)^2] + (\gamma - 1)\zeta\theta(\lambda - \zeta)}{(\lambda - \zeta)^2 - \gamma\theta} = \frac{2N(1 - \theta) - (\gamma - 1)\theta\zeta'}{\zeta - \lambda}. \tag{2.1.5}$$

Here $N = n/(n + 2)$.

As boundary conditions, we will use conditions (1.3.7), i.e. we will consider the problem of outflow of a gas into a vacuum. Conditions (1.3.6) must be satisfied on the disturbance front. Since a disturbance will propagate with respect to the gas at the speed of sound:

$$c_{\mathrm{f}}(t) = \sqrt{\gamma R T_{\mathrm{f}}} = \sqrt{\gamma(\gamma - 1)F_0}\, t^{n/2}, \tag{2.1.6}$$

it follows that

$$r_f(t) = \frac{2}{n+2} \sqrt{\gamma(\gamma-1)F_0}\, t^{1+\frac{n}{2}}$$

and hence, on the basis of (1.2.4) and (1.2.9),

$$\lambda_f = \sqrt{\gamma}. \tag{2.1.7}$$

If the gas is heat conducting, the speed of a disturbance is not known in advance and λ_f is determined in the course of solving the problem.

2. *Investigation of the disturbance front.* To the conditions on the front there corresponds in the (λ, ζ, θ) space the point

$$I(\lambda = \sqrt{\gamma},\ \zeta = 0,\ \theta = 1), \tag{2.2.1}$$

which is a singular point for the system of equations (2.1.4) and (2.1.5). The point I belongs to the family of singular points

$$(\zeta - \lambda)^2 - \gamma\theta = 0; \quad 2(\theta - 1) + \zeta(\lambda - \zeta) = 0 \tag{2.2.2}$$

and hence is not an isolated singular point.

To determine the behavior of the integral curves in a neighborhood of I, the linear terms are retained in the numerators and denominators of the right sides of (2.1.4) and (2.1.5):

$$\zeta' = N\frac{2\bar\theta + \sqrt{\gamma}\,\zeta}{2\sqrt{\gamma}(\bar\lambda - \zeta) - \gamma\bar\theta}, \qquad \bar\theta' = -\frac{N(1-\gamma)}{\sqrt{\gamma}}\frac{2\bar\theta + \sqrt{\gamma}\,\zeta}{2\sqrt{\gamma}(\bar\lambda - \zeta) - \gamma\bar\theta}, \tag{2.2.3}$$

$$\theta = 1 + \bar\theta, \quad \lambda = \sqrt{\gamma} + \bar\lambda,$$

after which (2.2.3) is easily integrated, giving us

$$\theta = 1 + \frac{\gamma-1}{\sqrt{\gamma}}\zeta, \tag{2.2.4}$$

$$\zeta = \frac{2 - \left(3 - \frac{2}{\gamma}\right)N}{\gamma+1}(\lambda - \sqrt{\gamma}) + C\zeta^{\frac{2}{N(3-2/\gamma)}}, \tag{2.2.5}$$

where C is an arbitrary constant.

Consequently, there passes through the point I a one-parameter family of integral curves forming a surface in the (λ, ζ, θ) space. In the approximation considered this surface is the plane with equation (2.2.4), and the family of integral curves is a pencil of curves with one and the same slope

$$\zeta'(\sqrt{\gamma}) = \begin{cases} \dfrac{2 - (3 - 2/\gamma)N}{\gamma+1}, & \text{if} & \dfrac{2}{N(3-2/\gamma)} > 1; \\[2ex] 0, & \text{if} & \dfrac{2}{N(3-2/\gamma)} \leqslant 1 \end{cases} \tag{2.2.6}$$

at I.

We note that $2/N(3 - 2/\gamma)$ is in a certain sense a critical value for the disturbance front: when $2/N(3 - 2/\gamma) \leqslant 1$ we have $\zeta'(\sqrt{\gamma}) = \theta'(\sqrt{\gamma}) = \delta'(\sqrt{\gamma}) = 0$, i.e. the solution together with its first derivatives is continuous on the front; when $2/N(3 - 2/\gamma) > 1$ the first derivatives are discontinuous.

3. *Infinite rate of dispersion.* We first note that when $N = 0$ we obtain a well-known problem on the adiabatic outflow into a vacuum of a gas having pressure $p_f = (\gamma - 1)F_0\rho_0$ at $t = 0$. In this case system (2.1.1)–(2.1.3) is easily integrated, and, by taking into account boundary conditions (1.3.6), (1.3.7) and (2.1.7), the solution is obtained in the form

$$\delta = \left(\frac{\gamma - 1}{\gamma + 1}\frac{\lambda}{\sqrt{\gamma}} + \frac{2}{\gamma + 1}\right)^{\frac{2}{\gamma - 1}}, \qquad \zeta = \lambda - \sqrt{\gamma}\,\delta^{\frac{\gamma - 1}{2}}, \qquad \theta = \delta^{\gamma - 1}. \qquad (2.3.1)$$

The rate of outflow of the gas is finite and equal to $\zeta_0 = -2\sqrt{\gamma}/(\gamma - 1)$.

We will prove that the rate of outflow ζ_0 is not finite in the general case $N > 0$.

We make a number of preliminary remarks. If the desired solution is assumed to be continuous, the quantity $\lambda - \zeta$ will vary within the limits

$$0 \leqslant \lambda - \zeta \leqslant \sqrt{\gamma} \qquad (2.3.2)$$

when $\lambda_0 \leqslant \lambda \leqslant \sqrt{\gamma}$ and vanish only when $\lambda = \lambda_0$. This follows from the fact that the condition $\lambda = \zeta = \text{const}$ can be satisfied only on a λ-line that is a trajectory. In the problem under consideration the boundary dispersing into a vacuum will be such a trajectory (see (1.3.3)). Condition (2.3.2) implies

$$\zeta'_0 \leqslant 1. \qquad (2.3.3)$$

We will also assume that the temperature θ_0 achieves its minimal value on the boundary, so that when $\lambda = \lambda_0$

$$\theta_0 < 1, \qquad (2.3.4)$$
$$\theta'_0 \geqslant 0, \ \theta''_0 > 0. \qquad (2.3.5)$$

In the sequel, when we obtain a solution, we will demonstrate that conditions (2.3.4) and (2.3.5) are satisfied.

We proceed to the proof. Suppose ζ_0 is finite. We will consider the two cases $\theta_0 \neq 0$ and $\theta_0 = 0$.

Suppose $\theta_0 \neq 0$. Then (2.1.5) together with (2.3.4) implies

$$\theta'_0 = \lim_{\lambda = \lambda_0 + 0} \theta'(\lambda) = \frac{2N(1 - \theta_0)\,\theta_0}{-\gamma\theta_0}\lim_{\lambda = \lambda_0 + 0}\frac{1}{\lambda - \zeta} = -\infty,$$

which contradicts (2.3.5).

Suppose $\theta_0 = 0$. Then (2.14) implies

$$\zeta'_0 = \lim_{\lambda = \lambda_0 + 0}\zeta'(\lambda) = -2N\lim_{\lambda = \lambda_0 + 0}\frac{1}{(\lambda - \zeta)^2 - \gamma\theta} = -\infty. \qquad (2.3.6)$$

We know $\zeta'_0 \neq +\infty$ from inequality (2.3.3). But (2.3.6) contradicts (2.3.5), since (2.1.5) implies

$$\theta'_0 = \lim_{\lambda = \lambda_0 + 0}\theta'(\lambda) = \frac{2N - (\gamma - 1)\lim\theta\zeta'}{\lim(\zeta - \lambda)} = -\infty.$$

Consequently the assumption of a finite rate of outflow is not valid, and one must put $\zeta = -\infty$.

4. *Expansion of the solution in a neighborhood of the gas-vacuum boundary.* We shall show that the temperature cannot vanish on the boundary with the vacuum when the rate of outflow is infinite. Moreover, its value is determined from the expansions obtained below.

Since

$$\lambda_0 = \zeta_0 = -\infty, \tag{2.4.1}$$

we have the inequality

$$\zeta_0' \geqslant 0. \tag{2.4.2}$$

Taking (2.3.3) into account, we get

$$0 \leqslant \zeta_0' \leqslant 1. \tag{2.4.3}$$

It follows from the latter inequality and (2.1.5) that

$$\theta' \approx \frac{2N}{\zeta - \lambda} < 0$$

when $\theta_0 = 0$. But this contradicts (2.3.5). Thus $\theta_0 \neq 0$.

To the boundary of the gas with the vacuum there corresponds the point

$$\text{II} \ (\lambda = -\infty, \ \ \zeta = -\infty, \ \ \theta_0 \neq 0), \tag{2.4.4}$$

which is a singular point for system (2.1.4), (2.1.5) and lies on the line of singular points

$$\lambda = -\infty, \quad \zeta = -\infty. \tag{2.4.5}$$

On the basis of (2.3.4) and (2.4.1) we have

$$\theta_0' = 0. \tag{2.4.6}$$

Equation (2.1.4) will therefore have the form

$$\zeta' = \frac{N\zeta \, (\lambda - \zeta)}{(\zeta - \lambda)^2 - \theta_0} \tag{2.4.7}$$

in a neighborhood of II. The point $\lambda = -\infty$, $\xi = -\infty$ is a singular point for (2.4.7). Let us apply the transformation

$$\zeta = \frac{1}{y}, \quad \lambda = \frac{1}{x} \tag{2.4.8}$$

and consider the behavior of the integral curves in the quadrant of interest to us, $y \leqslant 0$, $x \leqslant 0$. Equation (2.4.7) takes the form

$$y' = \frac{N \, (y - x) \, y^2}{x \, [(x - y)^2 - \theta_0 x^2 y^2]}, \tag{2.4.9}$$

while the singular point goes over into the origin. Its character is established on the basis of Frommer's results presented in §1.6.

In the present case the critical directions along which the integral curves can enter the singular point will be

$$1)\ u = 0; \quad 2)\ u = 1, \quad 3)\ u = \frac{1}{1-N}; \quad 4)\ u = \infty. \qquad (2.4.10)$$

The following qualitative deductions can be made in accordance with §1.6. Directions 1)–3) have a single integral curve (since $\partial \psi(u, 0)/\partial u < 0$ for $u = 0, 1, 1/(1-N)$), while direction 4) has an uncountable number of integral curves. A qualitative picture of the curves of equation (2.4.9) is presented in Figure 1.

The integral curve with slope 2) has the expansion

$$y = x - \frac{\theta_0}{N}\, x^3 + \frac{1+3N}{N^3}\, \theta_0^2 x^5 - \dots \qquad (2.4.11)$$

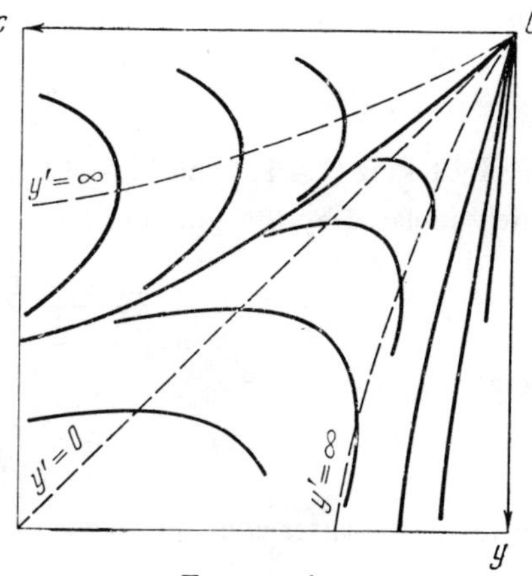

FIGURE 1

We note that the equation

$$y' = \frac{N\,(y-x)}{-\theta_0 x^3},$$

which is obtained from (2.4.9), can be integrated:

$$y = \frac{N}{\theta_0}\exp\left(\frac{N}{2\theta_0}\frac{1}{x^2}\right)\int_0^x \exp\left(-\frac{N}{2\theta_0}\frac{1}{t^2}\right)\frac{dt}{t^2}. \qquad (2.4.12)$$

The integral here is essentially the complementary error function [15], which has an asymptotic expansion about $x = 0$. Expansion (2.4.11) should therefore be used with caution, it being preferable to apply the representation of an integral curve in the form (2.4.12). The function (2.4.12) can be calculated with the use of the tables [15].

The asymptotic expansion (2.4.11) is an expansion of the desired solution since it represents the only integral curve leaving the origin and belonging to the domain $y - x > 0$ (see (2.3.2)).

Let us study the behavior of the integral curves of (2.1.5) by using expansion (2.4.11) for the function $\zeta(\lambda)$. We apply the transformation (2.4.8) to (2.1.5) and get

$$\frac{d\theta}{dx} = \frac{1}{xy}\frac{2N\,(1-\theta)\,y^2 - (\gamma - 1)\,\theta x^2 y'}{y - x}. \qquad (2.4.13)$$

This equation reduces on the basis of (2.4.11) to the equation

$$\frac{d\theta}{dx} = \frac{2N\,(1-\theta_0) - (\gamma - 1)\,\theta_0}{-\dfrac{\theta_0}{N}\,x^3} \qquad (2.4.14)$$

in a neighborhood of II. It follows from (2.4.6) that the desired solution satisfies the condition

$$\left(x^2\frac{d\theta}{dx}\right)_{x=0} = 0,$$

36　　　　　　　　　　　　　　　V. E. NEUVAŽAEV

which is satisfied only when the retained principal term of the numerator in (2.4.13) vanishes, i.e. when

$$2N(1-\theta_0)-(\gamma-1)\theta_0=0. \tag{2.4.15}$$

Thus

$$\theta_0=\frac{2N}{2N-1+\gamma}. \tag{2.4.16}$$

By taking (2.4.11) and (2.4.16) into account, equation (2.4.13) can be approximately replaced by the equation

$$\bar{\theta}'=\frac{\dfrac{2N(\gamma-1)}{2N-1+\gamma}x^2+\dfrac{(2N+\gamma-1)^2}{2}\bar{\theta}}{x^3}, \tag{2.4.17}$$

where

$$\bar{\theta}=\frac{2N}{2N-1+\gamma}-\theta.$$

Integrating the latter equation, we get

$$\theta=\frac{2N}{2N-1+\gamma}-\frac{2N(\gamma-1)}{2N-1+\gamma}\exp\left[\frac{(2N+\gamma-1)^2}{4x^2}\right]$$
$$\times\left[C+\int_0^x\exp\left(-\frac{(2N+\gamma-1)^2}{4x^2}\right)\frac{dx}{x}\right]. \tag{2.4.18}$$

Here C is an arbitrary constant.

Consequently, through the singular point

$$\mathrm{II}\left(\lambda=-\infty,\ \zeta=-\infty,\ \theta=\frac{2N}{2N-1+\gamma}\right), \tag{2.4.19}$$

of system (2.14), (2.15) corresponding to the boundary of the gas with the vacuum there passes the one-parameter family of integral curves described by (2.4.11) and (2.4.18), in the former of which θ_0 must be replaced by the right side of (2.4.16). In the (λ,ζ,θ) space this family forms a surface that is approximately representable by expansion (2.4.11) and on which the divergent pencil of integral curves (2.4.18) emanates from point (2.4.19).

5. *Overdeterminateness of the problem. Numerical integration of system* (2.1.4), (2.1.5). Thus one seeks a solution of the system of equations (2.14), (2.15) that joins points I and II. Despite the singular nature of these points, the boundary value problem turns out to be overdetermined.

For in order to choose the solution from the one-parameter family of curves emanating from I it is necessary to impose only one condition at $\lambda=-\infty$, for example, the condition $\zeta=-\infty$. If the problem is to have a solution, the remaining condition (2.4.16) must be automatically satisfied. The situation is analogous for the curves emanating from II.

The behavior of the integral curves in the (λ,ζ,θ) space is qualitatively described on the basis of the results of §§2.2 and 2.4 as follows. The manifold of integral curves emanating from I forms an integral surface (2.2.4), which, if a solution of the problem exists, necessarily passes through II. The manifold of integral curves emanating from II also forms an integral surface (2.4.11), (2.4.16), which must pass through I. Consequently, the desired solution belongs to the intersection of the integral surfaces (2.2.4) and (2.4.11), (2.4.16).

As we see, the behavior of the integral curves of system (2.1.4), (2.1.5) is such that a numerical integration of it will be unstable both from $\lambda = \sqrt{\gamma}$ to $\lambda = -\infty$ and in the opposite direction, from $\lambda = -\infty$ to $\lambda = \sqrt{\gamma}$. If system (2.1.4), (2.1.5) is written in the form

$$\frac{d\lambda}{d\zeta} = \frac{(\lambda - \zeta)^2 - \gamma\theta}{N\,[2\,(\theta - 1) + \zeta\,(\lambda - \zeta)]}\,, \tag{2.5.1}$$

$$\frac{d\theta}{d\zeta} = \frac{2\,(1 - \theta)\,[\theta - (\lambda - \zeta)^2] + (\gamma - 1)\,\zeta\theta\,(\lambda - \zeta)}{(\lambda - \zeta)\,[2\,(\theta - 1) + \zeta\,(\lambda - \zeta)]}\,, \tag{2.5.2}$$

it can be shown that an integration of (2.5.1) from $\zeta = -\infty$ to $\zeta = 0$ will be stable under a certain assignment of the function $\theta(\zeta)$, while an integration of (2.5.2) from $\zeta = 0$ to $\zeta = -\infty$ is stable for a certain function $\lambda(\zeta)$.

In fact, the assignment of a function $\theta(\zeta)$ in the space considered means the fixing of a certain surface. If this were the integral surface formed by the curves of I, a calculation on this surface to I would be stable since I is projected onto this surface from a point. Although the exact equation of this surface is not known, it can be determined approximately by (2.2.4) and (2.4.16). Under such an assignment of $\theta(\zeta)$ the character of the singular point I is retained, and an integration of (2.5.1) from $\zeta = -\infty$ to $\zeta = 0$ will be stable.

For a departure from II we use expansion (2.4.11), (2.4.16). As a result, we obtain a function $\lambda(\zeta)$ that will satisfy the boundary conditions $\lambda(-\infty) = -\infty$ and $\lambda(0) = \sqrt{\gamma}$. In our space this function will approximately describe the integral surface defined by the curves of II. Equation (2.5.2) has two singular points on this surface: point II is projected from a center while point I has the single integral curve defined by expansion (2.2.4). Departing from I by this expansion, we arrive at II.

We have described how a first approximation of the desired solution is obtained. The boundary conditions I and II are satisfied exactly in this connection. Higher approximations are obtained analogously.

A numerical integration shows that the indicated iterative process converges quite rapidly.

We note that the solution constructed in this way satisfies the boundary condition (1.3.7). For it follows from (2.1.2) that

$$\delta_0 = \lim_{\lambda = -\infty} \delta\,(\lambda) = \lim_{\lambda = -\infty} \exp\left(-\int_{\lambda}^{1} \frac{\zeta'\,d\lambda}{\lambda - \zeta}\right) = 0,$$

since the integral diverges when $\lambda = -\infty$.

6. *The case $\gamma = 1$.* Let us consider the special case of the problem when $\gamma = 1$. System (2.5.1), (2.5.2) takes the form

$$\frac{d\lambda}{d\zeta} = \frac{(\lambda - \zeta)^2 - \theta}{N\,[2\,(\theta - 1) + \zeta\,(\lambda - \zeta)]}\,, \tag{2.6.1}$$

$$\frac{d\theta}{d\zeta} = \frac{2\,(1-\theta)\,[\theta - (\lambda-\zeta)^2]}{(\lambda-\zeta)\,[2\,(\theta-1)+\zeta\,(\lambda-\zeta)]}\,. \tag{2.6.2}$$

The boundary conditions become

$$\text{I. } (\lambda = 1,\ \zeta = 0,\ \theta = 1),$$
$$\text{II. } (\lambda = -\infty,\ \zeta = -\infty,\ \theta = 1). \tag{2.6.3}$$

The plane

$$\theta = 1 \tag{2.6.4}$$

will be an integral surface([1]) on which equation (2.6.1) has the form

$$\frac{d\lambda}{d\zeta} = \frac{(\lambda-\zeta)^2 - 1}{N\zeta\,(\lambda-\zeta)}\,. \tag{2.6.5}$$

The iterative process indicated in §2.5 degenerates here into a single iteration: a numerical integration of (2.6.5) from the singular point ($\lambda = -\infty$, $\zeta = -\infty$) to the singular point ($\lambda = 1$, $\zeta = 0$), which is a saddle point.

We note that the problem of isothermal dispersion of an energy releasing gas reduces to the case $\gamma = 1$, which is also a limiting case for the problem of §3.

7. *The problem in Lagrangian coordinates.* The outflow of a gas into a vacuum has been considered in Lagrangian coordinates in [5]. In this case one is also able to reduce the study of a system of three first-order equations to the investigation of a system of two equations and indicate a stable direction of integration for it.

We replace λ in (1.2.4) by

$$\alpha = \frac{(n+2)\,m}{2\rho_0\,\sqrt{RB_0}\,t^{1+n/2}} \tag{2.7.1}$$

and use (1.2.4) to transform (1.1.5) when $\nu = \kappa = 0$ into the system of ordinary equations

$$\delta' = \frac{N\delta^2\,[\alpha\zeta + 2\,(\pi-\delta)]}{\alpha\,(\alpha^2 - \gamma\pi\delta)}\,, \tag{2.7.2}$$

$$\zeta' = \frac{N\,[\alpha\zeta + 2\,(\pi-\delta)]}{\alpha^2 - \gamma\pi\delta}\,, \tag{2.7.3}$$

$$\pi' = \frac{N\,[\gamma\zeta\pi\delta + 2\alpha\,(\pi-\delta)]}{\alpha^2 - \gamma\pi\delta}\,, \tag{2.7.4}$$

where $\pi = \theta\delta$.

If one introduces the new variables

$$\Pi = \frac{\pi}{\alpha}\,, \qquad \Delta = \frac{\delta}{\alpha}\,, \tag{2.7.5}$$

the problem reduces to the system of two equations

$$\begin{aligned}
\frac{d\Pi}{d\zeta} &= \frac{N\,[\gamma\zeta\Pi\Delta + 2\,(\Pi-\Delta)] - \Pi\,(1-\gamma\Pi\Delta)}{N\,[\zeta + 2\,(\Pi-\Delta)]}\,, \\[2mm]
\frac{d\Delta}{d\zeta} &= \frac{N\Delta\,[\zeta + 2\,(\Pi-\Delta)] - \Delta\,(1-\gamma\Pi\Delta)}{N\,[\zeta + 2\,(\Pi-\Delta)]}
\end{aligned} \tag{2.7.6}$$

with boundary conditions

([1]) This fact was pointed out by S. I. Sokolov.

$$\zeta = -\infty, \quad \Pi = -\infty, \quad \Delta = -\infty, \qquad (2.7.7)$$

$$\zeta = 0, \quad \Pi = \sqrt{\bar{\gamma}}, \quad \Delta = \sqrt{\bar{\gamma}}. \qquad (2.7.8)$$

Both of these conditions are nonisolated singular points of system (2.7.6), through each of which there passes a one-parameter family of integral curves having a common slope. Therefore, as in the case of Eulerian coordinates, a numerical integration of the system will be unstable in any direction.

A study of the behavior of the integral curves permits one to indicate new variables

$$\theta = \frac{\Pi}{\Delta}, \quad \Pi = \Pi$$

such that (2.7.6) reduces to the equations

$$\frac{d\theta}{d\zeta} = \frac{\theta\,[\zeta\,(\gamma\Pi - \theta) + 2\,(\theta - 1)\,(1 - \Pi)]}{\zeta\theta + 2\Pi\,(\theta - 1)}, \qquad (2.7.9)$$

$$\frac{d\Pi}{d\zeta} = \frac{N\,[\gamma\zeta\Pi^2 + 2\Pi\,(\theta - 1)] - \Pi\,(\theta - \gamma\Pi^2)}{N\,[\zeta\theta + 2\Pi\,(\theta - 1)]}. \qquad (2.7.10)$$

An integration of (2.7.9) from $\zeta = 0$ to $\zeta = -\infty$ is stable if $\Pi^0(\zeta)$ is assigned in accordance with (2.7.7) and (2.7.8). An integration of (2.7.10) from $\zeta = -\infty$ to $\zeta = 0$ is stable for the appropriate $\theta^0(\zeta)$. Therefore here again we can construct an iterative process.

§3. Outflow of a gas into a vacuum under a power law variation of the temperature on the boundary

1. *Statement of the problem.* Consider a heat-conducting gas with a coefficient of thermal conductivity given according to the law (1.2.2) and without energy release. The automodel equations for this case have the form

$$\delta\left[\frac{n}{n+2}\zeta + (\zeta - \lambda)\,\zeta'\right] + (\delta\theta)' = 0,$$

$$-\lambda\delta' + (\delta\zeta)' + \frac{\nu\delta\zeta}{\lambda} = 0, \qquad (3.1.1)$$

$$\frac{2n}{n+2}\left(\frac{\zeta^2}{2} + \frac{\theta}{\gamma - 1}\right) + (\zeta - \lambda)\left(\frac{\zeta^2}{2} + \frac{\theta}{\gamma - 1}\right)' + \frac{1}{\lambda^\nu\delta}\left[\lambda^\nu\left(\delta\theta\zeta - \frac{n+2}{2}A\frac{\theta^{(1+n)/n}}{\delta^l}\theta'\right)\right]' = 0.$$

We will study the plane problem of outflow of a gas into a vacuum when the temperature is given on the boundary in accordance with the law (1.3.2). The disturbance front propagates with a finite speed for $n > 0$. It can always be assumed that the quantity B_0 in (1.2.4) has been chosen so that $\lambda = 1$ on the front, in which case

$$\text{a) } \theta = 0; \quad \text{b) } \zeta = 0; \quad \text{c) } \delta = 1; \quad \text{d) } \theta^{(1+n)/n}\theta' = 0. \qquad (3.1.2)$$

Conditions (3.1.2) imply the continuity of the hydrodynamic quantities and heat flow on the distrubance front. Condition d) follows from condition a) if $|\theta'(1)| < \infty$; but, in general, this condition is not always satisfied.

It will be shown below that in the problem under consideration the gas flows out with an infinite speed, so that conditions (1.3.7) take the form

$$\lambda = \zeta = -\infty, \quad \delta = 0. \tag{3.1.3}$$

Thus a boundary value problem with conditions (3.1.2) and (3.1.3) on the right and left boundaries respectively is posed for the system of ordinary equations (3.1.1).

2. *The disturbance front is a singular point of system* (3.1.1). Let us investigate the disturbance front for any ν. System (3.1.1) can be reduced to the four first-order equations

$$\theta' = -\frac{\delta^l q}{A \lambda^\nu \theta^{(1+n)/n}}, \tag{3.2.1}$$

$$\zeta' = \frac{\left(\theta' + \dfrac{n}{n+2}\zeta\right)(\lambda - \zeta) + \dfrac{\nu\theta\zeta}{\lambda}}{(\zeta - \lambda)^2 - \theta}, \tag{3.2.2}$$

$$\delta' = \frac{\delta\left(\zeta' + \dfrac{\nu\zeta}{\lambda}\right)}{\lambda - \zeta}, \tag{3.2.3}$$

$$q' = -\frac{2\lambda^\nu\delta}{(2+n)(\gamma-1)}\left[(\gamma-1)\theta\left(\zeta' + \dfrac{\nu\zeta}{\lambda}\right) + \dfrac{2n}{n+2}\theta + (\zeta-\lambda)\theta'\right]. \tag{3.2.4}$$

The point

$$\mathrm{I}\,(\lambda = 1,\ \theta = 0,\ \zeta = 0,\ \delta = 1,\ q = 0), \tag{3.2.5}$$

corresponding to the disturbance front is a singular point and belongs to the manifold of singular points $\theta = 0,\ q = 0$. By retaining the principal terms in the right sides of system (3.2.1)–(3.2.4), we can find the integral curves which pass through I.

System (3.2.1)–(3.2.4) is replaced in a neighborhood of I by the system

$$\theta' = -\frac{q}{A\theta^{(1+n)/n}}, \tag{3.2.6}$$

$$\zeta' = \theta' + \frac{n}{n+2}\zeta, \tag{3.2.7}$$

$$\delta' = \delta(\zeta' + \nu\zeta), \tag{3.2.8}$$

$$q' = \frac{2}{(n+2)(\gamma-1)}\left(\theta' - \frac{2n}{n+2}\theta\right). \tag{3.2.9}$$

Equations (3.2.6) and (3.2.9) can be integrated independently of the others. After dividing the latter by the former, we get

$$\frac{dq}{d\theta} = \frac{2}{(n+2)(\gamma-1)}\frac{\dfrac{2n}{n+2}A\theta^{(2n+1)/n} + q}{q}. \tag{3.2.10}$$

Equation (3.2.10) is investigated by following Frommer (see §1.6). At the point $q = 0$, $\theta = 0$ it has two critical directions:

$$u = q/\theta = \frac{2}{(n+2)(\gamma-1)}, \tag{3.2.11}$$

$$u = q = 0. \tag{3.2.12}$$

Direction (3.2.11) has a single integral curve since $\partial\psi(u, 0)/\partial u \equiv -1 < 0$. Direction (3.2.12) is singular (see [14], §7), since the denominator $Q(\theta, q) = q$ in (3.2.10) vanishes when $q = 0$. It, too, has only one integral curve, the leading term in an expansion in increasing powers of θ of which is

$$q = -\frac{2n}{n+2} A\theta^{(2n+1)/n} + \cdots .$$

The behavior of the integral curves of (3.2.10) is depicted in Figure 2.

The desired solution must be found in the quadrant $\theta \geqslant 0$, $q \geqslant 0$ since (i) the dimensionless temperature θ is positive, and (ii) $q \geqslant 0$ on the basis of (3.2.1) and the

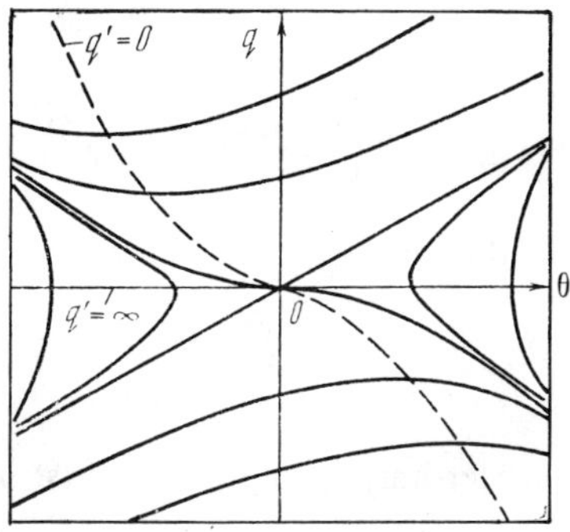

FIGURE 2

fact that $\theta'(0) < 0$. Consequently the solution will be the unique integral curve with slope (3.2.11):

$$q = \frac{2}{(n+2)(\gamma-1)}\theta + \cdots . \qquad (3.2.13)$$

The leading terms in the expansions of the remaining functions can be uniquely determined with the use of (3.2.13) from equations (3.2.6)–(3.2.8) and boundary conditions (3.2.5):

$$\theta = k(1-\lambda)^{n/(n+1)},$$

$$\zeta = \frac{n}{n+1}\,k\exp\left(\frac{n}{n+2}\lambda\right)\int_1^\lambda \exp\left(-\frac{n}{n+2}\lambda\right)\cdot(1-\lambda)^{-1/(n+1)}\,d\lambda,$$

$$\delta = \exp\left(\zeta + \nu\int_1^\lambda \zeta\,d\lambda\right), \qquad k = \left[\frac{2(n+1)}{A(n+2)(\gamma-1)n}\right]^{n/(n+1)} . \qquad (3.2.14)$$

Thus the singular point I has just one integral curve emanating from it and lying in the domain of interest to us.

Formulas (3.2.13) and (3.2.14) approximately describe the solution in the vicinity of the disturbance front. We note that the derivative of temperature is infinite (equal to zero) on the front when $n/(n+1) < 1$ $(n/(n+1) > 1)$. In the above problem $n/(n+1) < 1$ by virtue of the fact that $n > 0$.

3. *Rate of dispersion of a heat-conducting gas.* It is known that a gas without heat conduction flows out with the finite speed $-2c_0/(\gamma-1)$, where c_0 is the initial speed of sound in the gas at rest. It was proved in the preceding subsection that a gas flows out with infinite speed under the action of energy release. In the problem under consideration a heat-conducting gas also disperses at an infinite rate. Let us prove this. We will show that $\delta(\lambda_0) \neq 0$ if $\zeta_0 = \lambda_0$ is a finite quantity.

From (3.2.2) and (3.2.3) when $\nu = 0$ it follows that

$$(\ln\delta)' = \frac{\theta' + \dfrac{n}{n+2}\zeta}{(\lambda-\zeta)^2 - \theta} \equiv f(\lambda).$$

Integrating this equation over an interval $\lambda_0 \leqslant \lambda \leqslant \lambda_1$, we get

$$\delta(\lambda) = \delta(\lambda_1) \exp\left[-\int_{\lambda}^{\lambda_1} f(\lambda)\, d\lambda\right]. \tag{3.3.1}$$

We split the integral in (3.3.1) into two integrals:

$$J(\lambda) = -\int_{\lambda}^{\lambda_1} f(\lambda)\, d\lambda = -\frac{n}{n+2}\int_{\lambda}^{\lambda_1}\frac{\zeta\, d\lambda}{(\lambda-\zeta)^2-\theta} - \int_{\lambda}^{\lambda_1}\frac{\theta'\, d\lambda}{(\lambda-\zeta)^2-\theta}$$

and consider $\lim_{\lambda\to\lambda_0+0} J$. The first integral is bounded for $\lambda \to \lambda_0$. To see that the second integral is also bounded, we note that in a neighborhood of λ_0 it approximates the integral

$$-\int_{\lambda_0}^{\lambda_1}\frac{\theta'\, d\lambda}{(\lambda-\zeta)^2-\theta} = C_2 + \int_{\lambda_0}^{\lambda_1}\frac{\theta'}{\theta}\, d\lambda = C_2 + \ln\frac{\theta(\lambda_1)}{\theta(\lambda_0)};$$

where C_2 is a bounded constant. Consequently the integral

$$J(\lambda_0) = \lim_{\lambda=\lambda_0+0} J(\lambda) = C_1 + C_2 + \ln\frac{\theta(\lambda_1)}{\theta(\lambda_0)}$$

is bounded and $\delta(\lambda_0) \neq 0$, which contradicts (3.1.3). Therefore the rate of dispersion cannot be finite.

We note that the proof holds for any l.

If it is assumed that $\zeta_0 = -\infty$, one can indicate an integral curve for which $\delta_0 = 0$. This will be done in §3.5.

4. *Degeneracy of the solution for* $l \leqslant 0$. An expression for dimensionless flow is given by (3.2.1), which can be rewritten in the form

$$q = -\frac{A\theta^{(1+n)/n}}{\delta^l}\,\theta'.$$

According to its physical meaning, the flow $q_0 = q(-\infty)$ on a boundary flying straight off to infinity must be finite and positive ($0 < q_0 < \infty$). Therefore

$$\theta'(-\infty) = 0 \qquad \text{when} \qquad l > 0;$$
$$\theta'(-\infty) < 0,\ \neq -\infty \qquad \text{when} \qquad l = 0;$$
$$\theta'(-\infty) = -\infty \qquad \text{when} \qquad l < 0.$$

The fact that $\theta'(-\infty) \neq 0$ when $l \leqslant 0$ and $\theta(\lambda) \geqslant 0$ for $-\infty < \lambda \leqslant 1$ implies $\theta_0 = \infty$. Consequently the problem posed in §3.1 can be meaningful only if the temperature of the vacuum is assumed to be infinitely large. This means that under the power law variation of temperature (1.3.2) the problem has a degenerate solution when T_0 is finite and $l \leqslant 0$. A boundary dispersing at an infinite rate confines the flow of heat to infinity, and no motion takes place in this case.

5. *Behavior of the solution in a neighborhood of the gas-vacuum point.* To the conditions on a dispersing boundary there corresponds the point

$$\text{II}\ (\lambda = -\infty,\ \theta = \theta_0,\ \zeta = -\infty,\ \delta = 0,\ q = q_0), \tag{3.5.1}$$

where θ_0 and q_0 are subject to determination. Like point I, point II is a singular point for system (3.2.1)–(3.2.4) and belongs to the manifold of singular points $\lambda = -\infty$, $\zeta = -\infty$, $\delta = 0$.

From conditions (3.5.1) we know that $\theta'_0 = 0$ when $l > 0$. Therefore, when $\nu = 0$ equation (3.2.2) is approximated by the equation

$$\zeta' = \frac{\dfrac{n}{n+2}\,\zeta\,(\lambda - \zeta)}{(\lambda - \zeta)^2 - \theta_0}.$$

The latter has already been encountered and investigated in detail in §2.4. In the domain of interest to us it has a single integral curve emanating from the point $\lambda = -\infty$, $\zeta = -\infty$ and having the expansion (2.4.11).

It can be shown by making use of expansion (2.4.11) and expression (3.3.1) for the numerical density that $\lim_{\lambda=-\infty}\delta(\lambda) = 0$. In fact, the integrand $f(\lambda)$ in the improper integral of (3.3.1) grows unboundedly when $\lambda \to \infty$:

$$f(\lambda) \sim - \frac{n}{(n+2)\,\theta_0}\lambda.$$

Therefore $\delta(\lambda) \to 0$ as $\exp[-n\lambda^2/(n+2)\theta_0]$.

6. *A limit case* $(A = \infty)$. The solutions of the ordinary equations (3.2.1)–(3.2.4) depend on the numerical parameter A given by (1.2.8) and determined by the initial conditions of the problem. To each concrete problem there corresponds a specific value of A. The parameter A can vary from 0 to ∞. For fixed γ, R, ρ_0 and B_0 its variation is linearly connected with the coefficient κ_0, a decrease or increase of which corresponds to a decrease or increase in the coefficient of thermal conductivity. Therefore when $A = \infty$ we can put

$$\theta'(\lambda) = 0 \tag{3.6.1}$$

and consider equations (3.2.2) and (3.2.3) for $\nu = 0$. This case corresponds to an infinite conduction of heat and has already been considered in the problem with energy release (§2.6). There is no heat front, since heat instantly dissipates throughout the gas.

We can take $B_0 = T_0$ in (1.2.4) and (1.3.2), so that (3.6.1) implies $\theta(\lambda) = 1$ for all $\lambda \in (-\infty, \infty)$. Equation (2.6.5) is solved under the following boundary conditions: $\zeta = -\infty$ for $\lambda = -\infty$ and $\zeta = 0$ for $\lambda = +\infty$. The point $\lambda = \infty$, $\zeta = 0$ is an isolated singular saddle point of equation (2.6.5). The only admissible integral curve passing through this point is the coordinate axis $\zeta = 0$.

Thus $\zeta = 0$, i.e. the gas is at rest, for $1 \leqslant \lambda < \infty$.

The point $\lambda = 1$, $\zeta = 0$ is a singular point of nodal type and corresponds to a weak discontinuity. The solution in the interval $(-\infty, 1)$ is found by the numerical method presented in §2.6.

Thus the solution for $A = \infty$ can be qualitatively described by the following model: a heat wave instantly warms the whole gas, the boundary of the gas with the vacuum flies off to infinity for $t > 0$, and a rarefaction wave propagates at the isothermal speed of sound $\sqrt{RT_0}\,t^{n/2}$ with respect to the gas at rest.

7. *Construction of the solution for* $A \neq \infty$. *Statement for an isothermal jump.* When $A \neq \infty$ the problem reduces to the determination of a solution satisfying the system of ordinary equations (3.2.1)–(3.2.4) and the boundary conditions (3.1.3) and (3.2.5). The problem is solved numerically.

Conditions (3.2.5) on the disturbance front uniquely determine the integral curve with asymptotic behavior (3.2.13), (3.2.14). A numerical integration shows that there does not exist a continuous solution satisfying conditions (3.1.3) on the left boundary. Under an integration from $\lambda = 1$ to $\lambda = -\infty$ one encounters a point λ^0, ζ^0 at which $d\lambda/d\zeta = 0$ and $d^2\lambda/d\zeta^2 > 0$ (Figure 3). This singularity does not permit one to construct a continuous solution, and so a

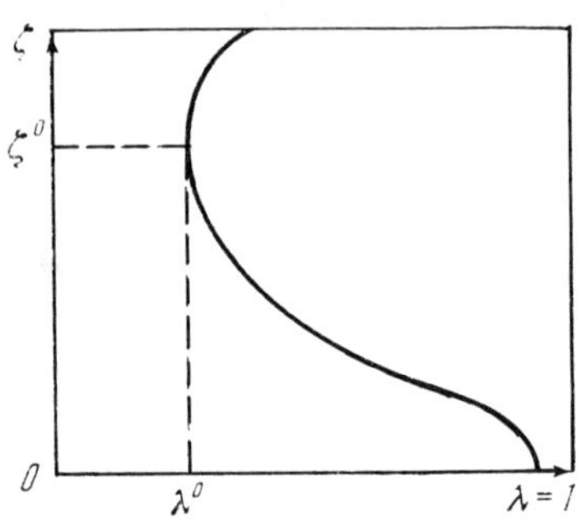

FIGURE 3

discontinuous solution is constructed with the use of relations (1.4.3) on a discontinuity.

In the statement of a problem with a discontinuity there appears a free parameter, say λ_1, since the three conditions on a discontinuity connect the four unknown quantities $\lambda_1, \zeta_2, \delta_2$ and q_2. This makes it possible to satisfy one of the conditions (3.1.3), viz, the condition $\zeta = -\infty$ for $\lambda = -\infty$. The condition $\delta_0 = 0$, as was shown in §3.5, is automatically satisfied.

8. *Numerical integration of equations* (3.2.1)–(3.2.4). The numerical integration of (3.2.1)–(3.2.4) from $\lambda = 1$ to $\lambda = \lambda^0$ is conducted by the usual pointwise method: the derivatives are replaced by differences and the solution is successively determined from $\lambda = \lambda_i$ to $\lambda = \lambda_i - h$, where h is a given mesh width in the coordinate λ. A calculation over the interval $[\lambda^0, 1]$ in the indicated direction does not lead to a loss of accuracy.

The difficulties noted in §1.5 arise in a numerical integration over the interval $(-\infty, \lambda_1]$. In contrast to the problem of §2, an isothermal discontinuity at $\lambda = \lambda_1$ is not a singular point of system (3.2.1)–(3.2.4). But when $A \to \infty$ the latter tends to a weak discontinuity, which is a singular point. Therefore the following iterative process is constructed for large A by analogy with the iterative method of §2: equation (3.2.2) is integrated for a given function $\theta(\lambda)$ from $\lambda = -\infty$ to $\lambda = \lambda_1$ with the use of expansion (2.4.11); then (3.2.1), (3.2.3) and (3.2.4) are integrated for the resulting function $\zeta(\lambda)$ from $\lambda = \lambda_1$ to $\lambda = -\infty$, equation (3.2.3) being solved with the use of (3.3.1). The process is repeated until convergence. It results in the satisfaction of conditions (3.1.3) on the left boundary, while on the right boundary for $\lambda = \lambda_1$ a value of ζ_2^0 is obtained that is generally not equal to the known quantity ζ_2. Therefore λ_1 is selected so that the equality $\zeta_2^0 = \zeta_2$ holds.

9. *The case of arbitrary A. Two classes of solutions.* With one method presented above, when $n > 0$ and $\nu = 0$ one is able to find only one class of solutions, which corresponds to large values of A. It can be characterized by the fact that on the interval $(-\infty, \lambda_1]$ the numerical temperature $\theta(\lambda)$ varies monotonically and achieves its minimal value at $\lambda = \lambda_1$. To this class there correspond for fixed γ, n and l the values of A lying in a certain interval $[A_*, \infty)$. The value A_* corresponds to the solution for which, in addition to $\theta(\lambda_1)$ being the minimal value of $\theta(\lambda)$ on the interval $(-\infty, \lambda_1]$, the condition $\theta'(\lambda_1) = 0$ is satisfied, i.e. the flow $q(\lambda_1) = 0$.

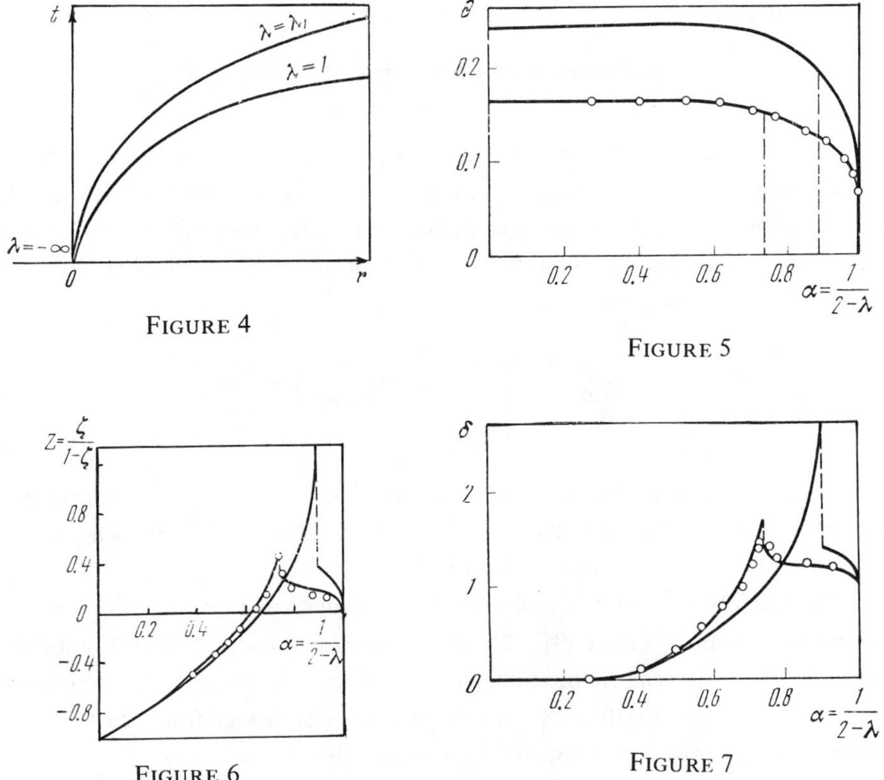

FIGURE 4

FIGURE 5

FIGURE 6

FIGURE 7

When $A < A_*$ there is reason to assume the existence of another class of solutions in which the temperature θ achieves its minimal value on the interval $(-\infty, \lambda_1]$ at an interior point λ_* at which $\theta'(\lambda_*) = 0$ and hence the flow $q(\lambda_*) = 0$. We have not been able to construct a solution of this class by the method presented above.[2]

Thus, if $A > A_*$, the solution can be qualitatively described as follows: when $t > 0$ there commences to propagate, with respect to the gas at rest, a disturbance (Figure 4) on whose front all of the quantities T, ρ, u, κ and $\partial T/\partial x$ vary continuously and only their derivatives are discontinuous. Such a disturbance is called a heat front (Figure 4, $\lambda = 1$). Behind the heat front there propagates a strong isothermal discontinuity ($\lambda = \lambda_1$), while the left boundary disperses with infinite speed into the vacuum ($\lambda = -\infty$). When $A = \infty$ the strong discontinuity goes over into a weak discontinuity, and when $A = 0$ the solution turns out to be degenerate, i.e. no motion occurs in this case.

A perfect gas with $\gamma = 1.4$ and coefficient of thermal conductivity $\kappa = \kappa_0 T^6/\rho^3$ was taken as an example. Two cases were investigated: 1) $A = 0.714 \cdot 10^6$ and 2) $A = 0.714 \cdot 10^5$. The solutions obtained both belong to the first type, when $A \geqslant A_*$ (Figures 5–7). A solution of the second type was constructed in [7] by the difference method of [11] for $A = 0.625$ and $\kappa = \kappa_0 T/\rho$.

In Figures 5–7 the approximate solutions obtained by the method of [11] are

---

(2) Such solutions have been constructed for the "piston" problem by a difference method in [6] and [7].

marked by small circles, the exact solutions are represented by a solid line and the discontinuities are indicated by a dashed line.

§4. Propagation of a spherical blast wave in a heat-conducting gas

1. *Preliminary remarks.* The automodel equations (3.1.1) contain an arbitrary parameter n characterizing the law of variation of the physical variables with time (see (1.2.4)). If one considers the problem without the energy release (1.2.3) and assumes that $n < 0$, one can indicate for each ν a value of n for which the total energy of the system remains constant with time [1]:

$$\begin{aligned}
n &= -6/5 \quad \text{for} \quad \nu = 2; \\
n &= -1 \quad \text{for} \quad \nu = 1; \\
n &= -2/3 \quad \text{for} \quad \nu = 0.
\end{aligned} \qquad (4.1.1)$$

The problem of an intense point explosion in which heat conduction is taken into account was investigated for cylindrical motions under a constant coefficient of thermal conductivity ($\kappa = \kappa_0$) by I. O. Bežaev in [16].

The case of spherical waves when $l = 0$ was considered by V. P. Korobeĭnikov [8] and later by L. A. Elliott [9]. The disturbance front considered by Korobeĭnikov is a shock wave with temperature jump and heat flow. The statement of the problem is different in our investigation. We assume that the disturbance front is a heat wave with continuous hydrodynamic quantities and heat flow (see (3.2.1)).

In the interior of the heat wave there exists an isothermal jump.

2. *Statement of the problem.* As was pointed out by G. M. Bam-Zelikovič [17], when $\nu = 2$ and $n = -6/5$ the energy equation in (3.1.1) has the integral

$$\frac{2}{5} A \frac{\theta^{1/6}}{\delta^l} \theta' - \delta \left(\frac{\zeta^2}{2} + \frac{\theta}{\gamma - 1}\right)(\zeta - \lambda) - \delta\theta\zeta = c. \qquad (4.2.1)$$

The constant c is determined from conditions (3.1.2):

$$c = 0. \qquad (4.2.2)$$

By taking (4.2.1) and (4.2.2) into account, system (3.1.1) for $\nu = 2$ can be written in the form

$$\theta' = \frac{5}{2} \frac{\delta^{c+1}}{A\theta^{1/6}} \left[\left(\frac{\zeta^2}{2} + \frac{\theta}{\gamma - 1}\right)(\zeta - \lambda) + \theta\zeta\right], \qquad (4.2.3)$$

$$\zeta' = \frac{\left(\theta' - \dfrac{3}{2}\zeta\right)(\lambda - \zeta) + \dfrac{2\theta\zeta}{\lambda}}{(\zeta - \lambda)^2 - \theta}, \qquad (4.2.4)$$

$$\delta' = \delta \frac{\zeta' + 2\zeta/\lambda}{\lambda - \zeta}. \qquad (4.2.5)$$

The boundary conditions for system (4.2.3)–(4.2.5) will be conditions (1.3.8) at the center and conditions a)–c) of (3.1.2) on the disturbance front (condition d) was used to determine the constant of integration c).

The physical parameter B_0 is connected with the initial parameters of the problem by the relation

$$B_0 = \frac{1}{R} \left(\frac{\alpha E_0}{\rho_0} \right)^{2/5}, \tag{4.2.6}$$

where α is defined by

$$\frac{1}{\alpha} = 4\pi \left(\frac{5}{2} \right)^3 \int_0^\ \delta \left(\frac{\zeta^2}{2} + \frac{\theta}{\gamma - 1} \right) \lambda^2 d\lambda. \tag{4.2.7}$$

3. *Behavior of the solution in a neighborhood of the center.* Condition (1.3.8) at the center determines a manifold of singular points for the system (4.2.3)–(4.2.5). It can be investigated if one takes into account the fact that in the problem under consideration $\delta(0) \neq 0$ and $\theta(0) < \infty$ (in contrast to the corresponding problem for adiabatic motion [1], where $\delta(0) = 0$ and $\theta(0) = \infty$). Then (4.2.3) implies

$$\theta' \approx \frac{5}{2} \frac{\delta_0^{\ell+1}}{A\theta_0^{1/6}} \frac{\theta_0}{\gamma - 1} (\gamma\zeta - \lambda). \tag{4.3.1}$$

It follows from (4.3.1) that (4.2.4) can be replaced in a neighborhood of the center by the equation

$$\zeta' = \frac{1}{\lambda} \left[\frac{5}{2} \frac{\delta_0}{A\,(\gamma - 1)\,\theta_0^{1/6}} \lambda^3 - 2\zeta \right]. \tag{4.3.2}$$

Integrating the latter equation, we get

$$\zeta = c\lambda^{-2} + \frac{\delta_0}{2A\theta_0^{1/6}\,(\gamma - 1)} \lambda^3. \tag{4.3.3}$$

The behavior of the integral curves is depicted in Figure 8.

The cubic parabola

$$\zeta = \frac{\delta_0}{2A\theta_0^{1/6}\,(\gamma - 1)} \lambda^3 \tag{4.3.4}$$

corresponds to the desired solution and represents the leading term of an expansion of it. The following behavior of the functions $\theta(\lambda)$ and $\delta(\lambda)$ can be determined with the use of (4.3.4):

$$\theta = \theta_0 - \frac{\zeta\delta_0\theta_0^{2/3}}{4\,(\gamma - 1)\,A} \lambda^2, \qquad \delta = \delta_0 + \frac{\zeta}{6} \frac{\delta_0^2}{(\gamma - 1)\,A\theta_0^{1/6}} \lambda^2. \tag{4.3.5}$$

Relations (4.3.4) and (4.3.5) were obtained in [8].

4. *Nonuniqueness of the solution under a discontinuity in the temperature.* The solution is constructed by the numerical method presented in §3.7. An isothermal jump (1.4.3) is between the disturbance front and the center.

One can drop the isothermicity assumption and, following [8], suppose that the temperature is discontinuous across the jump. There would then appear in conjunction with the free parameter λ_1 another free parameter, due to the fact that under a discontinuous temperature the three conditions on the jump connect five unknown quantities: $\lambda_1, \theta_2, \zeta_2, \delta_2$ and q_2.

Condition (1.3.8) at the center determines only one parameter, leaving the other, say λ_1, arbitrary. As a result, the solution of the problem is not unique: for each

$\lambda_1 \in (\lambda^0, 1]$ it is possible to construct an integral curve satisfying all of the boundary conditions. The solution for $\lambda_1 = 1$ was found in [8].

4a. *Two classes of solutions for ne= − 6/5.* The essential difference between the present problem and the problem of §3 is that here the exponent $n = -6/5 < -1$, so that on the basis of (3.2.14) the temperature on the disturbance front has a zero tangent, whereas when $n > 0$ the heat front is infinitely steep. The value of n also determines the power of the temperature on which the coefficient of thermal conductivity depends: the power $m < 1$ when $n < -1$, whereas $m > 1$ when $n > 0$.

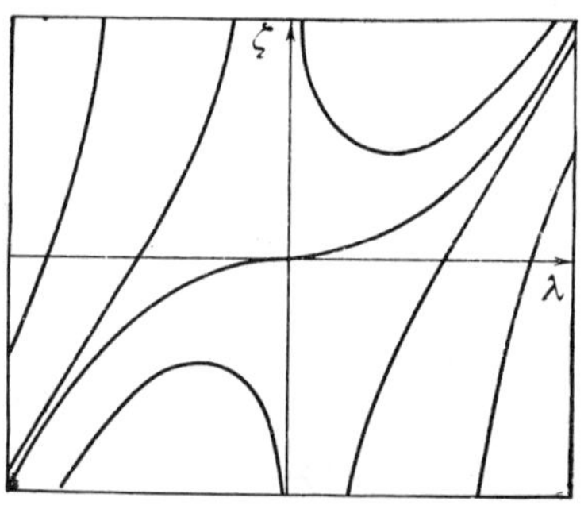

FIGURE 8

And yet for the problem under consideration the numerical parameter A, in the same way as for the problem of §3, determines two types of solutions. They can be separated in the following way: one contains all of the solutions that do not have singular points between the center and the isothermal jump, so that all of the functions in this group vary smoothly; the other contains the solutions having two singular points such that the functions ζ and δ behave smoothly at the point near the jump but are discontinuous at the other point.

The solutions are obtained without singularities for all $0 \leqslant A < A_*$. As $A \to 0$, the solution tends to the well-known solution of L. I. Sedov for an adiabatic explosion. The latter can be obtained from the equations under consideration by putting $A = 0$ and taking

$$\theta_1(1) = \frac{2(\gamma-1)}{(\gamma+1)^2}; \qquad \zeta_2(1) = \frac{2}{\gamma+1}; \qquad \delta_2(1) = \frac{\gamma+1}{\gamma-1}$$

as the conditions on the discontinuity $\lambda_1 = 1$. Sedov's solution is a limit solution for the present problem and is of the first type.

The second type is obtained when $A_* \leqslant A < \infty$. When $A = \infty$ the solution degenerates into the trivial solution. For the remaining A of the type considered there arise points at which the numerator and denominator in (4.2.4) vanish simultaneously. The denominator

$$\Delta = (\zeta - \lambda)^2 - \theta$$

in the terms of the original problem has the form

$$\left(u - \sqrt{RB_0}\, t^{-3/5\lambda}\right)^2 - RT,$$

so that its vanishing means that the corresponding point is moving at the isothermal speed of sound $\sqrt{RT}$ with respect to the gas.

The behavior of the solution is qualitatively represented by the (r, t) graphs in Figure 9 for $0 \leqslant A < A_*$ ($\lambda = 0$ is the center, $\lambda = \lambda_1$ is the isothermal jump and $\lambda = 1$ is the disturbance front) and in Figure 10 for $A_* \leqslant A < \infty$ ($\lambda = 0$ is the center, $\lambda = \lambda_3$

is a weak discontinuity, $\lambda = \lambda_1$ is the isothermal jump and $\lambda = 1$ is the disturbance front)

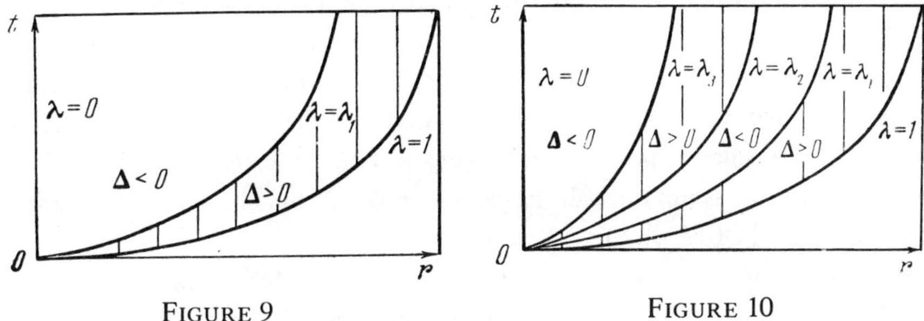

FIGURE 9   FIGURE 10

5. *Numerical integration of system* (4.2.3)–(4.2.5). The method of numerical integration of (4.2.3)–(4.2.5) is carried over bodily from §3.8. It is obvious (see Figure 8) that the behavior of the integral curves near the center is such that an integration of equation (4.2.4) for a given function $\theta(\lambda)$ will be stable from $\lambda = 0$ to $\lambda = \lambda_1$. An integration of (4.2.3) in the opposite direction (from $\lambda = \lambda_1$ to $\lambda = 0$), as a calculation shows, does not lead to an essential increase in the errors, and hence is always suitable in practice.

The case $A_* \leqslant A < \infty$ involves significant difficulties; the determination of the function $\zeta(\lambda)$ is complicated by the singular points of equation (4.2.4) arising in this case. For large A, however, the numerical temperature $\theta(\lambda)$ varies little for $\lambda \in [0, \lambda_1]$, so that it is possible to approximately assume that

$$\theta'(\lambda) = 0. \tag{4.5.1}$$

Then (4.2.4) takes the form

$$\zeta' = \frac{-\dfrac{3}{2}\zeta(\lambda - \zeta) + \dfrac{2\theta_1\zeta}{\lambda}}{(\zeta - \lambda)^2 - \theta_1}. \tag{4.5.2}$$

A similar equation is studied in [3]. It has the following singular points for the values of λ and ζ of interest to us: $O(0, 0)$, $B(\sqrt{\theta_1}, 0)$ and $C(4\sqrt{\theta_1}/3, \sqrt{\theta_1}/3)$.

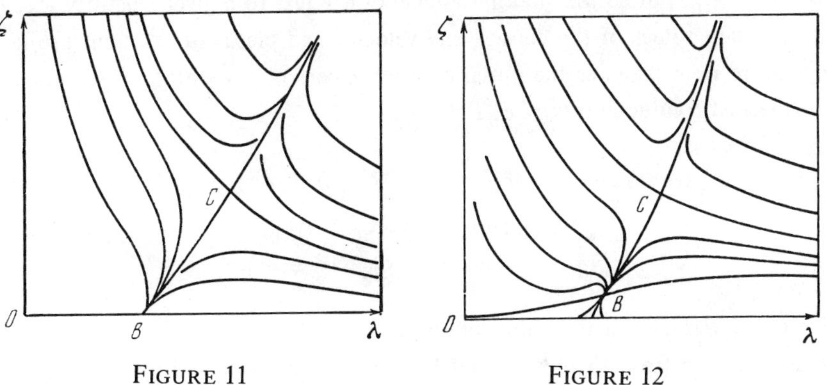

FIGURE 11   FIGURE 12

The singular point O is a saddle point. Through it pass two integral curves: the straight lines $\zeta = 0$ and $\lambda = 0$. Point B is a nodal point and C is a saddle point. The field of integral curves is depicted in Figure 11. The solution at each iteration will be a curve passing through the singular points O, B and C.

In those cases when one cannot assume that condition (4.5.1) is satisfied, equation (4.2.4) is studied for a suitable function $\theta(\lambda)$. As our examples show, this approach preserves the character of the singular points and hence the behavior of the integral curves. For a known $\theta(\lambda)$ one takes as the solution $\zeta(\lambda)$ the integral curve of (4.2.4) passing through the singular points B and C (Figure 12). We note that one could choose an integral curve not passing through the singular point C, but then one cannot satisfy the condition $\zeta_2^0 = \zeta_2$ (see §3.8).

A perfect gas with $\gamma = 1.4$ was taken as an example. The parameter A was set equal to 0.361, 0.721, 1.45 and 9.07. It turned out that $0.721 < A_* < 1.45$.

6. *Comparison with numerical solution.* As is well known, such a comparison is achieved by introducing an artificial viscosity that permits one to conduct a calculation uninteruptedly, without making special provision for discontinuities.

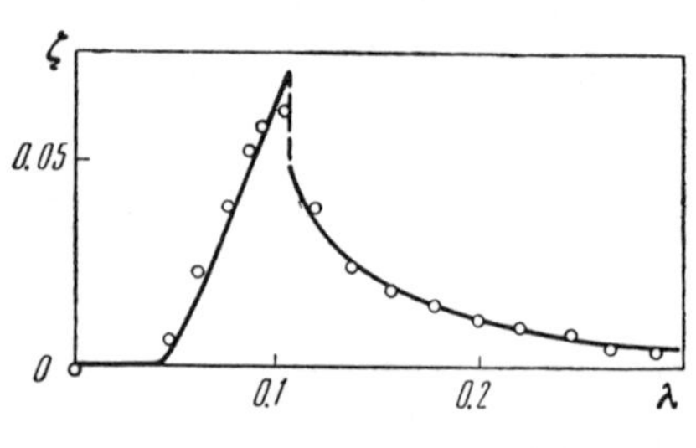

FIGURE 13

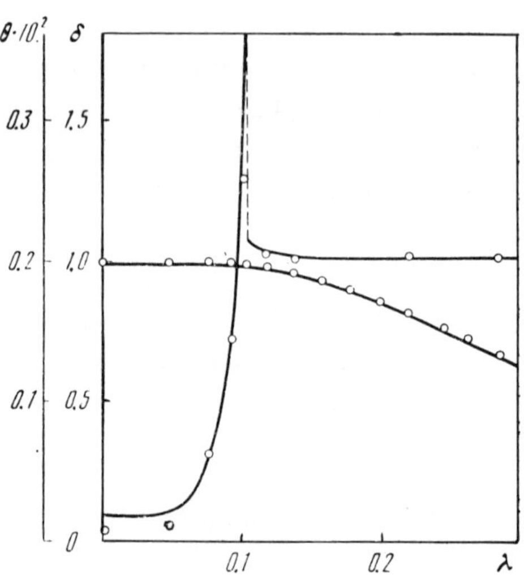

FIGURE 14

The solution constructed here can be obtained by the difference scheme proposed in [11]. To this end the initial conditions are given quite arbitrarily in a certain finite sphere of radius R_0, but so that the initial energy is equal to a given quantity E_0. The following constant values of the density and velocity and piecewise constant profile of the temperature were taken as the initial data for a perfect gas with $\gamma = 1.4$ and coefficient of thermal conductivity $\kappa < \kappa_0 T^{1/6}$:

$$\rho\,(0,\,r) = 1, \quad u\,(0,\,r) = 0, \quad 0 \leqslant r < \infty,$$

$$T\,(0,\,r) = \begin{cases} 0.186 \cdot 10^5, & 0 \leqslant r \leqslant 0.85 \cdot 10^{-2}, \\ 0, & 0.85 \cdot 10^{-2} < r < \infty. \end{cases} \qquad (4.6.1)$$

The value of κ_0 was chosen from the condition $A = 1.45$.

As can be seen from Figures 13 and 14, the solution obtained by the difference scheme of [11] is in good agreement with the exact solution (the approximate solutions obtained by the method of [11] are marked by small circles, the exact solutions are represented by a solid line and the discontinuities are indicated by a dashed line).

An automodel motion with a continuous temperature has the property that the solutions of nonautomodel problems with quite arbitrary initial conditions tend to it. As the initial data, one can choose any solution with a discontinuous temperature; for example, the solution obtained in [8]. In every case the motion reaches the automodel regime. In this sense solutions with a discontinuous temperature are unstable and go over into solutions with a continuous temperature.

BIBLIOGRAPHY

1. L. I. Sedov, *Similarity and dimensional methods in mechanics*, 4th ed., GITTL, Moscow, 1957; English transl., Academic Press, New York, 1959.   MR 19, 898; 21 #6840.

2. L. V. Ovsjannikov, *Group properties of differential equations*, Izdat. Sibirsk. Otdel. Akad. Nauk SSSR, Novosibirsk, 1962. (Russian) MR **26** #264.

3. V. P. Korobeĭnikov, N. S. Mel'nikova and E. V. Rjazanov, *Theory of point explosions*, Fizmatgiz, Moscow, 1961. (Russian)   MR **29** #5494.

4. L. Collatz, *Numerische Behandlung von Differentialgleichungen*, Die Grundlehren der math. Wissenschaften, Band 60, Springer-Verlag, Berlin, 1951; 3rd ed. (in English), 1960. MR 13, 285; 22 #322.

5. V. E. Neuvažaev, *Flow of a gas into a vacuum with energy release obeying a power law*, Dokl. Akad. Nauk SSSR 141 (1961), 1058–1060 = Soviet Phys. Dokl. 6 (1961/62), 1055–1057.

6. P. P. Volosevič et al., *The solution of the homogeneous plane problem of the motion of a piston in an ideal heat-conducting gas*, Ž. Vyčisl. Mat. i Mat. Fiz. 3 (1963), 159–169 = USSR Comput. Math. and Math. Phys. 3 (1963), 204–218.

7. V. E. Neuvažaev, *Discharge of gas into a vacuum with temperature at the boundary subject to an exponential law*, Prikl. Mat. Meh. 30 (1966), 1015–1021 = J. Appl. Math. Mech. 30 (1966), 1210–1217.

8. V. P. Korobeĭnikov, *On propagation of a strong spherical blast wave in a gas with heat conduction*, Dokl. Akad. Nauk SSSR 113 (1957), 1006–1009. (Russian)   MR 19, 956.

9. L. A. Elliott, *Similarity methods in radiation hydrodynamics*, Proc. Roy. Soc. London Ser. A **258** (1960), 287–301.   MR 22 #7530.

10. V. E. Neuvažaev, *The propagation of a spherical blast wave in a heat-conducting gas*, Prikl. Mat. Meh. 26 (1962), 1094–1099 = J. Appl. Math. Mech. 26 (1962), 1657–1665.

11. N. N. Janenko and V. E. Neuvažaev, *A method of computing gas-dynamic motions with nonlinear heat conduction*, Trudy Mat. Inst. Steklov. 74 (1966), 138–140 = Proc. Steklov Inst. Math. 74 (1967), 150–153.

12. I. M. Gel'fand and O. V. Lokucievskiĭ, *The double-sweep method for the solution of difference equations*, Appendix II to the book of S. K. Godunov and V. S. Rjaben'kiĭ, *Theory of difference schemes. An introduction*, Fizmatgiz, Moscow, 1962; English transl., North-Holland, Amsterdam; Interscience, New York, 1965.   MR 29 #724; 31 #5346.

13. V. V. Nemyckiĭ and V. V. Stepanov, *Qualitative theory of differential equations*, OGIZ, Moscow, 1947; English transl., Princeton Math. Ser., no. 22, Princeton Univ. Press, Princeton, N. J., 1960.   MR **10**, 612; 22 #12258.

14. M. Frommer, *Die Integralkurven einer gewöhnlichen Differentialgleichungen erster Ordnung in der Umbegung rationaler Unbestimmtheitsstellen*, Math. Ann. 99 (1928), 222–272.

15. V. N. Faddeeva and N. M. Terent'ev, *Tables of values of the functions $w(z) = e^{-z^2}(1 + 2i\pi^{-1/2}\int_0^z e^{t^2}\,dt)$ for a complex argument*, GITTL, Moscow, 1954; English transl., Math. Tables Ser., vol. 2, Pergamon Press, Oxford, 1961,   MR 16, 920; 22 #12740.

16. I. O. Bežaev, *On the influence of viscosity and thermal conductivity of a gas on the propagation of a strong explosion*, Teoret. Gidromeh., no. 11 (L. I. Sedov, editor), Oborongiz, Moscow, 1953. (Russian)

17. G. M. Bam-Zelikovič, *Propagation of strong blast waves*, Teoret. Gidromeh., no. 4 (L. I. Sedov, editor), Oborongiz, Moscow, 1949. (Russian)

PLANE POTENTIAL FLOWS WITH STATIONARY STREAMLINES

UDC 517.951

V. A. SUČKOV

ABSTRACT. In this paper the geometric properties of two-dimensional potential flows are studied. A class of flows with stationary streamlines is singled out from the entire set of motions. It is shown that such flows have well-defined analytic and geometric properties.

Bibliography: 5 items.

§1. Statement of the problem.
The equation of gas dynamics in a moving reference frame

Consider the equations of plane potential motion of a barotropic gas,

$$\frac{\partial u_i}{\partial t} + u_k \frac{\partial u_i}{\partial x_k} + \frac{\partial \varphi}{\partial x_i} = 0,$$

$$\frac{\partial \varphi}{\partial t} + u_k \frac{\partial \varphi}{\partial x_k} + c^2 \frac{\partial u_k}{\partial x_k} = 0, \qquad \frac{\partial u_1}{\partial x_2} = \frac{\partial u_2}{\partial x_1}, \qquad i,\, k = 1,\, 2. \tag{1.1}$$

Here the $u_i(x_1, x_2, t)$ are the components of the velocity vector, $\varphi(x_1, x_2, t)$ is the enthalpy and $c(x_1, x_2, t)$ is the speed of sound.

The streamlines of the flow are determined from the solution of the differential equation [1]

$$\frac{dx_1}{u_1(x_1,\, x_2,\, t)} = \frac{dx_2}{u_2(x_1,\, x_2,\, t)}. \tag{1.2}$$

They generally depend on the time t. Plane flows with stationary streamlines are characterized by the differential relation [2]

$$u_2 \frac{\partial u_1}{\partial t} - u_1 \frac{\partial u_2}{\partial t} = 0. \tag{1.3}$$

It is necessary to investigate the compatibility of the overdetermined system (1.1), (1.3) (see [3]). To this class of flows belong steady-state and nonsteady-state flows with plane or axial symmetry, where the stationary streamlines are rectilinear [1].

Substituting the functions v (the modulus of the velocity vector), θ (the inclination of the velocity vector with respect to the x_1 axis) and Δ such that

$$u_1 = v \cos \theta, \qquad u_2 = v \sin \theta, \qquad \Delta = \varphi + \frac{v^2}{2} \tag{1.4}$$

into (1.1) and making certain transformations, we obtain the system

$$\frac{\partial v}{\partial t} + \frac{\partial \Delta}{\partial s_1} = 0, \qquad v \frac{\partial \theta}{\partial t} + \frac{\partial \Delta}{\partial s_2} = 0, \qquad \frac{\partial v}{\partial s_2} = v \frac{\partial \theta}{\partial s_1},$$

$$\frac{\partial \Delta}{\partial t} + 2v \frac{\partial \Delta}{\partial s_1} - v^2 \frac{\partial v}{\partial s_1} + c^2 \left(\frac{\partial v}{\partial s_1} + v \frac{\partial \theta}{\partial s_2} \right) = 0. \tag{1.5}$$

AMS (MOS) subject classifications (1970). Primary 76G20.

Here s_1 and s_2 are the arc lengths of the streamlines and of the lines orthogonal to them respectively, $\partial f/\partial s_1$ is the derivative along a streamline and $\partial f/\partial s_2$ is a normal derivative with respect to a streamline:

$$\frac{\partial f}{\partial s_1} = \frac{\partial f}{\partial x_1}\cos\theta + \frac{\partial f}{\partial x_2}\sin\theta, \qquad \frac{\partial f}{\partial s_2} = -\frac{\partial f}{\partial x_1}\sin\theta + \frac{\partial f}{\partial x_2}\cos\theta, \qquad (1.6)$$

where f is any of the functions v, θ or Δ. The curvatures of these lines are respectively denoted by K_1 and K_2:

$$K_1 = \frac{\partial\theta}{\partial s_1}, \qquad K_2 = \frac{\partial\theta}{\partial s_2}. \qquad (1.7)$$

In order for the streamlines to be stationary it is necessary and sufficient that

$$\theta = \theta(x_1,\ x_2). \qquad (1.8)$$

We will consider flows with stationary streamlines in the curvilinear coordinates q_1 and q_2 such that

$$x_i = x_i(q_1,\ q_2), \qquad ds_i = H_i dq_i, \qquad K_i = \frac{(-1)^i}{H_1 H_2}\frac{\partial H_i}{\partial q_{3-i}}, \qquad (1.9)$$

where the $H_i = H_i(q_1, q_2)$, $i = 1, 2$, are the Lamé coefficients. Substituting (1.9) into (1.5) and putting here

$$\frac{\partial\theta}{\partial t} = 0, \qquad v = \frac{\psi}{H_1}; \qquad (1.10)$$

we obtain the equations for potential flows with stationary streamlines in the curvilinear coordinates q_1 and q_2:

$$\frac{\partial\psi}{\partial t} + \frac{\partial\Delta}{\partial q_1} = 0, \qquad \frac{\partial\psi}{\partial q_2} = 0, \qquad \frac{\partial\Delta}{\partial q_2} = 0, \qquad (1.11)$$

$$\frac{\partial\Delta}{\partial t} + 2\psi\frac{\partial\Delta}{\partial q_1}l - \psi^2\frac{\partial\psi}{\partial q_1}l^2 - \psi^3 m + c^2\left(l\frac{\partial\psi}{\partial q_1} + \psi n\right) = 0, \qquad (1.12)$$

where

$$l = \frac{1}{H_1^2}, \qquad m = \frac{l}{2}\frac{\partial l}{\partial q_1}, \qquad n = \frac{1}{2}\frac{\partial l}{\partial q_1} + \frac{l}{H_2}\frac{\partial H_2}{\partial q_1}. \qquad (1.13)$$

We confine ourselves to the cases of a polytropic gas. For it

$$c^2 = \varkappa(2\Delta - \psi^2 l). \qquad (1.14)$$

For an adiabatic flow $\kappa = (\gamma - 1)/2$; for an isothermal flow $\kappa = 0$ and $2\kappa\Delta = 1$.

REMARK 1. If the functions $H_i(q_1, q_2)$ are known, the functions $x_i(q_1, q_2)$ are determined from the system of equations

$$\left(\frac{\partial x_1}{\partial q_i}\right)^2 + \left(\frac{\partial x_2}{\partial q_i}\right)^2 = H_i^2, \qquad i = 1,\ 2,$$

$$\frac{\partial x_1}{\partial q_1}\frac{\partial x_1}{\partial q_2} + \frac{\partial x_2}{\partial q_1}\frac{\partial x_2}{\partial q_2} = 0. \qquad (1.15)$$

Equations (1.15) are compatible if H_1 and H_2 satisfy Gauss' relation [4]

$$\frac{\partial}{\partial q_1}\left(\frac{1}{H_1}\frac{\partial H_2}{\partial q_1}\right)+\frac{\partial}{\partial q_2}\left(\frac{1}{H_2}\frac{\partial H_1}{\partial q_2}\right)=0. \tag{1.16}$$

REMARK 2. The system of equations (1.11), (1.12), (1.16) is invariant relative to the transformation

$$q_i'=\int f_i(q_i)\,dq_i,\qquad H_i=H_i'f_i,\qquad \psi=\psi'f_1. \tag{1.17}$$

When applying transformation (1.17) below, we will omit the primes in denoting the transformed functions.

§2. Investigation of compatibility for steady motion

It will be assumed that the functions $H_i(q_1,q_2)$ are known and satisfy (1.16). Let us consider the compatibility of system (1.11), (1.12). The functions Δ and ψ depend only on q_1 and t, while the functions l, m and n depend only on q_1 and q_2. The variable q_2 enters into (1.12) only as a parameter through the functions l, m and n. We will obtain compatibility conditions on the functions $H_i(q_1,q_2)$ by differentiating (1.12) with respect to q_2. In what follows we will need to know the derivative

$$\frac{\partial c^2}{\partial q_2}=-\varkappa\psi^2 l_2. \tag{2.1}$$

Here and below a subscript on Δ, ψ, l, m or n denotes differentiation with respect to the corresponding variable; for example, $\Delta_1=\partial\Delta/\partial q_1$, $\psi_t=\partial\psi/\partial t$, $l_2=\partial l/\partial q_2$, etc.

To illustrate the method we first consider steady flow, where $\psi_t=\Delta_t=0$ [5]. From (1.11) we get $\Delta=\Delta_0=\text{const}$, while from (1.12) we obtain the following differential equation for $\psi(q_1)$:

$$-\psi^2\psi_1 l^2-\psi^3 m+c^2(\psi_1 l+\psi n)=0, \tag{2.2}$$

Differentiating with respect to q_2, we get

$$-2\psi^2\psi_1 ll_2-\psi^3 m_2+c^2(\psi_1 l_2+\psi n_2)-\varkappa\psi^2 l_2(\psi_1 l+\psi n)=0. \tag{2.3}$$

If $l_2=0$, we find from (1.16), (1.17) and (2.3) that

$$H_1=1,\qquad H_2=\alpha q_1+\beta,$$
$$K_1=0,\qquad K_2=\frac{\alpha.}{\alpha q_1+\beta}, \tag{2.4}$$

where α and β are constants. Equation (2.2) will have the form

$$-\psi^2\psi_1+c^2\left(\psi_1+\frac{\alpha}{\alpha q_1+\beta}\,\psi\right)=0. \tag{2.5}$$

If $\alpha=0$ and $\beta=1$, we have from (2.5) that $\psi=\psi_0=\text{const}$ and hence $v=v_0$ or, in other words, a constant motion. If $\alpha=1$ and $\beta=0$, we have $K_1=0$, $K_2=1/q_1$ and hence a source or sink type flow. The function $v(q_1)$ is determined from the differential equation

$$(c^2-v^2)\frac{dv}{dq_1}+\frac{c^2v}{q_1}=0,\qquad c^2=\varkappa(2\Delta_0-v^2). \tag{2.6}$$

THEOREM. *A steady plane potential flow with rectilinear streamlines is either a constant motion or a source or sink type motion.*

Suppose now $l_2 \neq 0$. Making the transformation (1.17) with $f_1 = \psi(q_1)$ and $f_2 = 1$, we get from (2.2) that

$$\varkappa \left(2\Delta_0 - l\right) n = m. \tag{2.7}$$

Integrating with respect to q_1, we get

$$H_2 = \left(2\varkappa\Delta_0 - \frac{\varkappa}{H_1^2}\right)^{-1/2\varkappa} H_1. \tag{2.8}$$

Here, in accordance with Remark 2, we have set an arbitrary function of q_2 equal to unity. For an isothermal flow we find, upon passing to the limit as $\kappa \to 0$ in (2.8), that

$$H_2 = H_1 \exp \frac{1}{2H_1^2}. \tag{2.9}$$

Let us substitute (2.8) into (1.16). We then obtain the following differential equation of second order for the function $H_1(q_1, q_2)$:

$$\frac{\partial^2 H_1}{\partial q_2^2} + \frac{H_2}{H_1} \frac{dH_2}{dH_1} \frac{\partial^2 H_1}{\partial q_1^2} = \Phi, \tag{2.10}$$

where the function $\Phi(q_1, q_2)$ does not contain second derivatives of H_1 and

$$\frac{dH_2}{dH_1} = \frac{H_2}{H_1}\left(1 - \frac{v^2}{c^2}\right). \tag{2.11}$$

Equation (2.10) will be hyperbolic (elliptic) for a supersonic (subsonic) motion. If $H_1(q_1, q_2)$ is determined from (2.10) with the use of (2.8) and (2.11), then $H_2(q_1, q_2)$, the speed v and the enthalpy ϕ can be found from (2.8) and the relations

$$v = \frac{1}{H_1}, \qquad \varphi = \Delta_0 - \frac{v^2}{2}. \tag{2.12}$$

§3. Nonsteady-state case. Solutions having a function-type arbitrariness

We proceed to a consideration of unsteady motion. Differentiating (1.12) with respect to q_2, we get

$$2\psi\Delta_1 l_2 - 2\psi^2\psi_1 l l_2 - \psi^3 m_2 + c^2 \left(\psi_1 l_2 + \psi n_2\right) - \varkappa\psi^2 l_2 \left(\psi_1 l + \psi n\right) = 0. \tag{3.1}$$

For rectilinear streamlines $l_2 = 0$. Completely analogously to the steady case, we find from (1.12), (1.16), (1.17) and (3.1) that

$$H_1 = 1, \qquad H_2 = \alpha q_1 + \beta, \qquad K_1 = 0, \qquad K_2 = \frac{\alpha}{\alpha q_1 + \beta};$$

$$\Delta_t + 2\psi\Delta_1 - \psi^2\psi_1 + c^2 \left(\psi_1 + \frac{\alpha\psi}{\alpha q_1 + \beta}\right) = 0, \tag{3.2}$$

$$c^2 = \varkappa\left(2\Delta - \psi^2\right).$$

If $\alpha = 0$ and $\beta = 1$, we get $H_1 = H_2 = 1$ and $K_1 = K_2 = 0$, i.e. a plane one-dimensional motion. If $\alpha = 1$ and $\beta = 0$, we get $H_1 = 1$, $H_2 = q_1$, $K_1 = 0$ and $K_2 = 1/q_1$, i.e. a cylindrical one-dimensional motion.

THEOREM. *An unsteady plane potential flow with rectilinear stationary streamlines is a motion with plane or axial symmetry.*

Let us consider the case when $l_2 \neq 0$. Dividing (3.1) by l_2 and differentiating once again with respect to q_2, we get

$$-2\psi^2\psi_1 l_2 - \psi^3 \frac{\partial}{\partial q_2}\left(\frac{m_2}{l_2}\right) + c^2\psi \frac{\partial}{\partial q_2}\left(\frac{n_2}{l_2}\right) - 2\psi^2\varkappa\,(\psi_1 l_2 + \psi n_2) = 0. \qquad (3.3)$$

Repeating this operation and dividing by ψ, we finally get

$$\psi^2 M = c^2 N, \qquad (3.4)$$

where

$$M = \frac{\partial}{\partial q_2}\left[\frac{1}{l_2}\frac{\partial}{\partial q_2}\left(\frac{m_2}{l_2}\right) + 3\varkappa\,\frac{n_2}{l_2}\right],$$
$$N = \frac{\partial}{\partial q_2}\left[\frac{1}{l_2}\frac{\partial}{\partial q_2}\left(\frac{n_2}{l_2}\right)\right]. \qquad (3.5)$$

We will separately investigate the two possible cases:
1) $M \neq 0$, $N \neq 0$;
2) $M = N = 0$.

In this section we investigate the first case. If $c^2 = 1$, equation (3.4) implies $\psi_t = 0$. Applying transformation (1.17) with $f_1 = \psi(q_1)$ and $f_2 = 1$ to system (1.11), (1.12), we find

$$\Delta = \alpha t + \beta, \qquad (3.6)$$
$$\alpha + n = m, \qquad (3.7)$$

where α and β are constants.

Thus, for isothermal flow in this case

$$v = \frac{1}{H_1}, \qquad \varphi = \alpha t + \beta - \frac{v^2}{2}, \qquad (3.8)$$

and the functions H_1 and H_2 are determined from the system of two differential equations (1.16) and (3.7). If $\alpha = 0$ and $\beta = \Delta_0$, we have the steady case already considered.

Suppose $c^2 = \varkappa(2\Delta - \psi^2 l)$. From (3.4) we have

$$\Delta = z\,(q_1)\,\frac{\psi^2}{2}, \qquad (3.9)$$

where

$$z\,(q_1) = l + \frac{M}{\varkappa N}.$$

From (3.9) we find the following derivatives of Δ:

$$\Delta_t = \psi\psi_t z, \qquad \Delta_1 = \psi\psi_1 z + z_1\frac{\psi^2}{2}. \qquad (3.10)$$

Substituting (3.9) and (3.10) into system (1.11), (1.12), we get

$$\psi_t + \psi\psi_1 z + \frac{\psi^2}{2}z_1 = 0, \qquad (3.11)$$

$$\psi_1(z - l)\,[(\varkappa + 1)l - z] + \psi\left[\varkappa n\,(z - l) - m - z_1\left(\frac{z}{2} - l\right)\right] = 0. \qquad (3.12)$$

Since the coefficient of ψ_1 in (3.12) is not equal to zero, we have

$$\psi = r\,(t)\,f\,(q_1), \qquad (3.13)$$

where $r(t)$ is an arbitrary function of integration and the function $f(q_1)$ satisfies the equation

$$(z - l)[(\varkappa + 1)l - z]\frac{df}{dq_1} + \left[\varkappa n(z - l) - m - z_1\left(\frac{z}{2} - l\right)\right]f = 0. \qquad (3.14)$$

Using (1.17) with $f_1 = f(q_1)$ and $f_2 = 1$, we get from (3.11)–(3.14) that

$$\psi = r(t), \qquad (3.15)$$

$$r_t + \frac{r^2}{2}z_1 = 0, \qquad (3.16)$$

$$\varkappa n(z - l) - m - z_1\left(\frac{z}{2} - l\right) = 0. \qquad (3.17)$$

From (3.16) for $r_t \neq 0$ we find

$$r = \frac{1}{t}, \qquad z = 2q_1. \qquad (3.18)$$

Substituting (3.9), (3.15) and (3.18) into (1.4), (1.10) and (3.17), we arrive at the following expressions for the speed and enthalpy:

$$v = \frac{1}{tH_1}, \qquad \varphi = \frac{1}{t^2}\left(q_1 - \frac{1}{2H_1^2}\right), \qquad (3.19)$$

the functions H_1 and H_2 being determined from the system of two differential equations (1.16) and

$$\varkappa n(2q_1 - l) - m - 2(q_1 - l) = 0. \qquad (3.20)$$

We have proved the following

THEOREM. *A plane potential flow with curvilinear stationary streamlines and function-type arbitrariness is either steady or belongs to the following class of solutions:*

1) $\varkappa = 0, \qquad v = \frac{1}{H_1}, \qquad \varphi = \alpha t + \beta - \frac{v^2}{2};$

2) $\varkappa \neq 0, \qquad v = \frac{1}{tH_1}, \qquad \varphi = \frac{1}{t^2}\left(q_1 - \frac{1}{2H_1^2}\right).$

The functions $H_1(q_1, q_2)$ and $H_2(q_1, q_2)$ are determined from the system of two differential equations (1.16) and (3.7) (when $\kappa = 0$) or (3.20) (when $\kappa \neq 0$).

§4. Determination of the time dependence of solutions having a constant-type arbitrariness

Let us investigate the second case $M = N = 0$. Integrating the differential equations

$$\frac{\partial}{\partial q_2}\left[\frac{1}{l_2}\frac{\partial}{\partial q_2}\left(\frac{m_2}{l_2}\right) + 3\varkappa\frac{n_2}{l_2}\right] = 0,$$

$$\frac{\partial}{\partial q_2}\left[\frac{1}{l_2}\frac{\partial}{\partial q_2}\left(\frac{n_2}{l_2}\right)\right] = 0 \qquad (4.1)$$

with respect to q_2 three times, we find

$$n = \Phi_1 l^2 + \Phi_2 l + \Phi_3,$$

$$m = F_1 l^2 + F_2 l + F_3 - \varkappa l\,(\Phi_1 l^2 + \Phi_2 l + \Phi_3), \qquad (4.2)$$

where the $\Phi_i = \Phi_i(q_1)$ and $F_i = F_i(q_1)$, $i = 1, 2, 3$, are arbitrary functions of integration. Substituting (4.2) into (1.12) and noting that the coefficients of the various powers of l must vanish, we get

$$\Delta_t - \psi^3 F_3 + 2\varkappa\Delta\psi\Phi_3 = 0,$$
$$2\psi\Delta_1 + 2\varkappa\Delta\left(\psi_1 + \psi\Phi_2\right) - \psi^3 F_2 = 0, \qquad (4.3)$$
$$2\varkappa\Delta\Phi_1 - \psi\psi_1\left(\varkappa + 1\right) - \psi^2 F_1 = 0.$$

We must consider the compatibility of the four first-order differential equations obtained by adjoining the first equation $\psi_t + \Delta_1 = 0$ of (1.11) to system (4.3), for the two functions Δ and ψ of two independent variables q_1 and t, as well as the compatibility of the system (1.16), (4.2) of three differential equations for the two functions H_1 and H_2 of two independent variables q_1 and q_2. In the latter system the two equations (4.2) are of first order while equation (1.16) is of second order. The compatibility conditions give certain relations for the functions Φ_i and F_i, $i = 1, 2, 3$. Assuming $\kappa + 1 \neq 0$, we equate the second-order mixed derivatives of $\psi(q_1, t)$ and obtain the following equation of third degree for the function $z = 2\kappa\Delta\psi^{-2}$:

$$A_3 z^3 + A_2 z^2 + A_1 z + A_0 = 0, \qquad (4.4)$$

where the coefficients A_i, $i = 0, 1, 2, 3$, are functions of q_1 only.

If not all of the coefficients $A_i(q_1)$ are equal to zero, then z is determined from (4.4) and hence is a function of q_1 only. This case was considered in §3.

Suppose now all of the A_i are equal to zero. From the equation $A_3 = 0$ we get

$$\Phi_1\left(q_1\right) = 0. \qquad (4.5)$$

Integrating the third equation of (4.3) with respect to q_1 under condition (4.5), we find

$$\psi = r\left(t\right) f\left(q_1\right),$$

where $r(t)$ is an arbitrary function of integration and

$$f\left(q_1\right) = \exp\left[-\frac{1}{\left(\varkappa + 1\right)} \int F_1 dq_1\right].$$

The transformation (1.17) with $f_1 = f(q_1)$ and $f_2 = 1$ gives

$$\psi = r\left(t\right), \qquad (4.6)$$

which implies

$$F_1\left(q_1\right) = 0. \qquad (4.7)$$

Integrating the first equation of (1.11) with respect to q_1, we get

$$\Delta = -r'\left(t\right) q_1 + s\left(t\right), \qquad (4.8)$$

where $r' = dr/dt$ and $s(t)$ is an arbitrary function.

When $\Phi_1 = F_1 = 0$ the compatibility conditions for system (1.11), (4.3) are

$$\Phi_2' = \varkappa\Phi_2^2, \qquad (4.9)$$
$$F_2' = \varkappa\Phi_2 F_2, \qquad (4.10)$$
$$\Phi_3' = -\frac{\Phi_2 F_2}{2}, \qquad (4.11)$$
$$F_3' = -\varkappa\Phi_2 F_3 + \varkappa F_2\Phi_3 - \frac{F_2^2}{2}. \qquad (4.12)$$

From the differential equations (4.9)–(4.12) we can determine the functions $\Phi_i(q_1)$ and $F_i(q_1)$, $i = 1, 2, 3$. The substitution of these functions into (4.3) leads to equations for determining the functions $r(t)$ and $s(t)$. Let us illustrate these remarks by considering the following two cases.

1) $\kappa = 0$. From equations (4.9)–(4.12) we have

$$\Phi_2 = k_2, \qquad \Phi_3 = -\frac{k_1 k_2}{2} q_1 + k_3,$$

$$F_2 = k_1, \qquad F_3 = -\frac{k_1^2}{2} q_1 + k_4. \tag{4.13}$$

Substituting (4.5)–(4.8) and (4.13) into (4.3), we get for the isothermal case

$$r' = -\frac{k_1}{2} r^2 + \frac{k_2}{2}, \qquad s' = -k_3 r + k_4 r^3. \tag{4.14}$$

2) $\kappa \neq 0$. This case breaks down into two subcases.

a) If $\Phi_2 = 0$,

$$F_2 = k_1, \qquad \Phi_3 = k_3, \qquad F_3 = \frac{k_1}{2}(2\kappa k_3 - k_1) q_1 + k_4 \tag{4.15}$$

and equations (4.3) imply

$$r' = -\frac{k_1}{2} r^2, \qquad s' + 2\kappa k_3 rs = k_4 r^3. \tag{4.16}$$

b) If $\Phi_2 \neq 0$,

$$F_2 = k_1 q, \qquad \Phi_2 = k_2 q, \qquad \Phi_3 = -\frac{k_1}{2\kappa} q + k_3,$$

$$F_3 = \frac{k_1 k_3}{k_2} - \frac{k_1^2}{2\kappa k_2} q + \frac{k_4}{q}, \tag{4.17}$$

where

$$q = \frac{1}{(1 - \kappa k_2 q_1)},$$

and equations (4.3) imply

$$r'' + 2\kappa k_3 rr' - \kappa k_2 k_4 r^3 = 0, \qquad 2\kappa k_2 s = 2r' + k_1 r^2. \tag{4.18}$$

The quantities k_1, k_2, k_3 and k_4 in (4.13)–(4.18) are constants.

§5. Investigation of the compatibility of the equations for solutions having a constant-type arbitrariness

Let us now consider the compatibility of the overdetermined system (1.16), (4.2). We write out this system after substituting (1.13), (4.5) and (4.7) into it:

$$\frac{l}{2} \frac{\partial l}{\partial q_1} = -\kappa \Phi_2 l^2 + (F_2 - \kappa \Phi_3) l + F_3,$$

$$\frac{1}{2} \frac{\partial l}{\partial q_1} + \frac{l}{H_2} \frac{\partial H_2}{\partial q_1} = \Phi_2 l + \Phi_3, \tag{5.1}$$

$$\frac{\partial}{\partial q_1}\left(\frac{1}{H_1} \frac{\partial H_2}{\partial q_1}\right) + \frac{\partial}{\partial q_2}\left(\frac{1}{H_2} \frac{\partial H_1}{\partial q_2}\right) = 0.$$

Letting $\omega = \sqrt{l H_2}$, we rewrite (5.1) in the form

$$\frac{1}{l} \frac{\partial l}{\partial q_1} = A, \tag{5.2}$$

$$\frac{1}{\omega} \frac{\partial \omega}{\partial q_1} = B, \tag{5.3}$$

$$\frac{\partial x}{\partial q_2} = xy + \omega^2 C, \tag{5.4}$$

where

$$A = -2\varkappa\Phi_2 + 2z\,(F_2 - \varkappa\Phi_3) + 2z^2F_3, \qquad z = \frac{1}{l},$$

$$(5.5)$$

$$B = \Phi_2 + z\Phi_3, \qquad C = \frac{\partial}{\partial q_1}\,(2B - A) + B\,(2B - A);$$

$$x = \frac{1}{l}\,\frac{\partial l}{\partial q_2}, \qquad y = \frac{1}{\omega}\,\frac{\partial\omega}{\partial q_2}. \tag{5.6}$$

From (4.9)–(4.12) and (5.5) we have

$$C = \sum_{k=0}^{4} C_k z^k, \tag{5.7}$$

where the coefficients C_k are the following functions of q_1 only:

$$C_0 = 2\,(\varkappa + 1)^2\,\Phi_2^2, \qquad C_1 = \Phi_2\,[-(3 + 7\varkappa)\,F_2 + 4\,(\varkappa + 1)^2\,\Phi_3],$$
$$C_2 = 5F_2^2 - 2\,(5\varkappa + 3)\,F_2\Phi_3 - 2\,(3\varkappa + 1)\,\Phi_2 F_3 + 2\,(\varkappa + 1)\,(2\varkappa + 1)\,\Phi_3^2, \quad (5.8)$$
$$C_3 = 6F_3\,[2F_2 - (2\varkappa + 1)\,\Phi_3], \qquad C_4 = 8F_3^2.$$

For the compatibility of system (5.2)–(5.4) we must have

$$x^2P = \omega^2Q. \tag{5.9}$$

Here

$$P = P_1 z + P_2 z^2, \qquad\qquad Q = \sum_{k=0}^{6} Q_k z^k, \tag{5.10}$$

$$P_1 = 2F_2 + (1 - 2\varkappa)\,\Phi_3, \qquad P_2 = 8F_3, \tag{5.11}$$

where the coefficients Q_k are homogeneous forms of third degree in the variables F_2, F_3, Φ_2 and Φ_3.

Equation (5.9) can be investigated by separately considering the following two cases (assuming $x \neq 0,\ \omega \neq 0$):

1) (5.9) is satisfied identically, in which case

$$P = Q = 0;$$

2) (5.9) is not an identity, in which case

$$PQ > 0.$$

For the first case we have

$$P_1 = P_2 = Q_k = 0, \qquad k = 0,\ 1,\ \ldots,\ 6. \tag{5.12}$$

From (4.9)–(4.12) and (5.12) we find

$$F_2 = \Phi_2 = F_3 = \Phi_3 = 0. \tag{5.13}$$

It therefore follows from (1.11) and (4.3) that $\Delta_t = \varphi_t = 0$; in other words, we have a steady flow, which was investigated in §2. It remains to consider the second case.

Differentiating (5.9) with respect to q_1 and q_2 and substituting there the values of the derivatives of the functions l, ω, x and y, we get

$$Q \sum_{k=1}^{2} kP_k z^k = P \sum_{k=0}^{6} kQ_k z^k - 2P^2C, \tag{5.14}$$

$$P \sum_{k=0}^{6} Q'_k z^k - Q \sum_{k=1}^{2} P'_k z^k = 2PQ\Big(l\,\frac{\partial A}{\partial l} - B\Big) + 2P^2AC. \tag{5.15}$$

Equalities (5.14) and (5.15) are rearrangements of respectively seventh and ninth degree polynomials in z whose coefficients depend only on q_1:

$$\sum_{k=0}^{7} A_k z^k = 0, \qquad \sum_{k=0}^{9} B_k z^k = 0. \tag{5.16}$$

Since the coefficients of these polynomials must vanish individually, we have a system of eighteen equations for the four functions F_2, F_3, Φ_2 and Φ_3:

$$A_k = 0, \quad k = 0, 1, \ldots, 7; \qquad B_k = 0, \quad k = 0, 1, \ldots, 9. \tag{5.17}$$

From the equations $A_0 = 0$ and $A_1 = 0$ we get

$$P_1 Q_0 = 0, \qquad P_2 Q_0 - P_1^2 C_0 = 0.$$

Since $P_1^2 + P_2^2 \neq 0$, it follows that $Q_0 = 0$, i.e.

$$\Phi_2 = 0. \tag{5.18}$$

On the basis of (4.13) and (4.15) we then have

$$F_2 = k_1, \quad F_3 = \frac{k_1}{2}(2\varkappa k_3 - k_1) q_1 + k_4, \qquad \Phi_3 = k_3. \tag{5.19}$$

From the equations $A_3 = 0$ and $B_6 = 0$ we obtain the equation

$$(k_3 - k_1)[(2\varkappa + 1)k_3 - k_1][(\varkappa + 1)k_3 - k_1] = 0, \tag{5.20}$$

which implies

$$k_1 = (1 + 2\varepsilon\varkappa) k_3, \qquad \varepsilon = 0, \ 1/2, \ 1. \tag{5.21}$$

where ϵ is equal to either 0, ½ or 1.

We calculate the functions $P(\epsilon)$ and $Q(\epsilon)$:

$$P(\varepsilon) = \frac{1}{l^2}[(4\varkappa\varepsilon - 2\varkappa + 3) k_3 l + 8 F_3],$$

$$Q(\varepsilon) = \frac{4\varepsilon(1 - \varepsilon)\varkappa^2 k_3^2 P}{l^2} - \frac{P}{l^4}[(4\varkappa\varepsilon - 2\varkappa + 1) k_3 l + 2 F_3]^2. \tag{5.22}$$

It follows that the condition $PQ > 0$ is not satisfied, i.e. system (5.1) is incompatible, whenever $\epsilon(1 - \epsilon)\varkappa k_3 = 0$. Therefore $\epsilon = \text{½}$, $\varkappa \neq 0$, $k_3 \neq 0$ and the condition $PQ > 0$ takes the form

where
$$(\varkappa^2 - 1)(w - w_1)(w - w_2) > 0, \tag{5.23}$$

$$w = -\frac{k_3 l}{F_3}, \qquad w_1 = -\frac{2}{\varkappa - 1}, \qquad w_2 = \frac{2}{\varkappa + 1}. \tag{5.24}$$

It can be verified that all of the remaining equations of (5.17) are satisfied identically under condition (5.21), i.e. when

$$k_1 = (\varkappa + 1) k_3. \tag{5.25}$$

The hyperbolas $w = w_1$, $w = w_2$ and the straight line $\varkappa = 1$ divide the halfplane $\varkappa > 0$ into three admissible domains:

1) $0 < \varkappa < 1, \ w_2 < w < w_1$;
2) $\varkappa > 1, \ w > w_2 > w_1$;
3) $\varkappa > 1, \ w < w_1 < w_2$.

§6. Integration of the equations for solutions having a constant-type arbitrariness

Substituting (5.5), (5.18), (5.19) and (5.25) into system (5.2), (5.3), we get

$$\frac{\partial l}{\partial q_1} = 2k_3 + \frac{2}{l}F_3, \qquad \frac{1}{\omega}\frac{\partial \omega}{\partial q_1} = \frac{k_3}{l}, \tag{6.1}$$

where $F_3 = (\kappa^2 - 1)k_3^2 q_1/2 + k_4$. In terms of $w = -k_3 l/F_3$, system (6.1) becomes

$$\frac{\partial w}{\partial q_1} = -\frac{k_3^2}{F_3 w}\left[2(w-1) + w^2\frac{F_3'}{k_3^2}\right], \tag{6.2}$$

$$\frac{1}{\omega}\frac{\partial \omega}{\partial q_1} = -\frac{k_3^2}{F_3 w}. \tag{6.3}$$

Let us first consider the case $\kappa = 1$. Then $F_3 = k_4$ and $F_3' = 0$. Integrating (6.2), we get

$$w + \ln(w-1) = -\frac{2k_3^2}{k_4}q_1 + F(q_2), \tag{6.4}$$

where $1 < w < \infty$, $k_3/k_4 < -H_1^2$ and $F(q_2)$ is an arbitrary function of integration. Integrating (6.3) and setting the arbitrary function of q_2 obtained thereby equal to unity (in accordance with Remark 2), we find

$$\omega = \sqrt{w-1}. \tag{6.5}$$

With the use of the derivative of (6.4) with respect to q_2, we get

$$x = \frac{1}{l}\frac{\partial l}{\partial q_2} = \frac{1}{w}\frac{\partial w}{\partial q_2} = \frac{F'(w-1)}{w^2}. \tag{6.6}$$

Substituting the expressions obtained for P, Q, x and ω into (5.9), we get $F' = \pm 2k_3^2/k_4$. It can be assumed that $k_3 = -k_4 = 1$. Then, taking the plus sign for F after setting the arbitrary constant of integration equal to zero ($F = 2q_2$), we get from (6.4) and (6.5) that

$$\frac{1}{H_1^2} + \ln\left(\frac{1}{H_1^2}-1\right) = 2(q_1 + q_2), \qquad H_1^2 + H_2^2 = 1. \tag{6.7}$$

Integrating (4.16), we find

$$r = \frac{1}{t}, \qquad s = -\frac{\ln t}{t^2}. \tag{6.8}$$

Suppose now $\kappa \neq 1$. We can set $k_4 = 0$. Then (6.2) and (6.3) can be written as follows:

$$\frac{\partial w}{\partial q_1} = -\frac{(w-w_1)(w-w_2)}{wq_1}, \qquad \frac{1}{\omega}\frac{\partial \omega}{\partial q_1} = -\frac{2}{(\kappa^2-1)wq_1}. \tag{6.9}$$

Integrating with respect to q_1, we get

$$q_1 = \frac{F(q_2)\,\omega}{\sqrt{|w-w_1||w-w_2|}}, \qquad \omega = \left|\frac{w-w_2}{w-w_1}\right|^{1/2\kappa}, \tag{6.10}$$

where

$$F(q_2) = \frac{2q_2}{\sqrt{|\kappa^2-1|}}. \tag{6.11}$$

Here we can set $k_1 = 2$. Then $k_3 = w_1$, and equations (4.16) give

$$r = \frac{1}{t}, \qquad s = s_0 t^{-4\varkappa/(\varkappa+1)}, \tag{6.12}$$

where s_0 is an arbitrary constant. The functions H_1 and H_2 are determined from the formulas

$$H_1^{-2} = (1 - \varkappa) w q_1, \qquad H_2 = H_1 \omega. \tag{6.13}$$

Thus we have proved the following

THEOREM. *The functions $v = (tH_1)^{-1}$ and $\Delta = q_1 t^{-2} + s$, where $s = -t^{-2} \ln t$ $(s = s_0 t^{-4\kappa/(\kappa+1)})$ and H_1 and H_2 are determined by (6.7) (by (6.13)) for $\kappa = 1$ $(\kappa \neq 1)$, describe an unsteady plane potential flow with curvilinear stationary streamlines and having a constant-type arbitrariness.*

§7. The results in Cartesian coordinates

Let us consider flows with stationary streamlines in the variables x_1, x_2, t. The assumption that the streamlines are stationary permitted us to obtain the time dependence of the solution in the variables q_1, q_2 and t explicitly. This dependence is the same in the variables x_1, x_2 and t, so that the solution must have the form

$$u_i = rV_i, \qquad \Delta = -r'V + s, \qquad i = 1, 2, \tag{7.1}$$

where r and s are known functions of t while the V_i and V are functions of x_1 and x_2 only. The functions u_1, u_2 and Δ satisfy the equations

$$\frac{\partial u_i}{\partial t} + \frac{\partial \Delta}{\partial x_i} = 0, \qquad \frac{\partial u_1}{\partial x_2} = \frac{\partial u_2}{\partial x_1}, \tag{7.2}$$

$$c^2 = \varkappa [2\Delta - (u_1^2 + u_2^2)],$$

$$\frac{\partial \Delta}{\partial t} + 2u_k \frac{\partial \Delta}{\partial x_k} - u_i u_k \frac{\partial u_i}{\partial x_k} + c^2 \frac{\partial u_k}{\partial x_k} = 0. \tag{7.3}$$

Equations (7.2) imply

$$V_i = \frac{\partial V}{\partial x_i}, \qquad i = 1, 2. \tag{7.4}$$

We will consider equation (7.3) for the dependencies $r = r(t)$ and $s = s(t)$ obtained above. When $\kappa = 0$, $r = 1$ and $s = \alpha t + \beta$ it follows from (7.3) that the function $V(x_1, x_2)$ must satisfy the equation

$$(V_i V_k - \delta_{ik}) V_{ik} = \alpha, \tag{7.5}$$

which coincides with the equation for the velocity potential of a steady isothermal flow when $\alpha = 0$. When $\kappa \neq 0$, $r = 1/t$ and $s = 0$ the function $V(x_1, x_2)$ must satisfy

$$[V_i V_k - \varkappa \delta_{ik} (2V - V_1^2 - V_2^2)] V_{ik} + 2 (V - V_1^2 - V_2^2) = 0. \tag{7.6}$$

The arbitrariness of the solutions of (7.5) and (7.6) consists of two functions of a single argument. When $\kappa = 1$, $r = 1/t$, $s = -t^{-2} \ln t$ and the functions H_1 and H_2 are determined from (6.7), it follows from (1.15) that

$$\frac{\partial x_1}{\partial q_1} = H_1 \cos \theta, \qquad \frac{\partial x_1}{\partial q_2} = -H_2 \sin \theta,$$
$$\frac{\partial x_2}{\partial q_1} = H_1 \sin \theta, \qquad \frac{\partial x_2}{\partial q_2} = H_2 \cos \theta,$$

(7.7)

where θ, viz. the inclination of the velocity vector with respect to the x_1 axis, is determined from the equations

$$\frac{\partial \theta}{\partial q_1} = -\frac{1}{H_2}\frac{\partial H_1}{\partial q_2}, \qquad \frac{\partial \theta}{\partial q_2} = \frac{1}{H_1}\frac{\partial H_2}{\partial q_1}.$$

(7.8)

Differentiating (6.7) with respect to q_1 and q_2, we get

$$\frac{\partial H_1}{\partial q_i} = \frac{2}{\Phi'}, \qquad \frac{\partial H_2}{\partial q_i} = -\frac{2H_1}{\Phi' H_2},$$

(7.9)

where

$$\Phi(H_1) = \frac{1}{H_1^2} + \ln\left(\frac{1}{H_1^2} - 1\right).$$

Substituting (7.9) into (7.8), we find

$$\frac{\partial \theta}{\partial q_1} = \frac{\partial \theta}{\partial q_2} = -\frac{2}{\Phi' H_2}.$$

(7.10)

The integration of (7.10) gives

$$\theta = \theta_0 + \text{arc} \cos H_1.$$

We can assume that $\theta_0 = 0$. Then

$$H_1 = \cos \theta, \qquad H_2 = \sin \theta.$$

(7.11)

Substituting $H_1 = \cos \theta$ into the first equation of (6.7), we obtain the following equation for determining $\theta(q_1, q_2)$:

$$\frac{1}{\cos^2 \theta} + \ln(\tan^2 \theta) = 2(q_1 + q_2).$$

(7.12)

Let us take q_1 and θ as the independent variables in (7.7). Then

$$\frac{\partial x_1}{\partial \theta} = -\frac{\sin \theta}{\cos^3 \theta}, \qquad \frac{\partial x_1}{\partial q_1} = 1,$$
$$\frac{\partial x_2}{\partial \theta} = \frac{1}{\cos^2 \theta}, \qquad \frac{\partial x_2}{\partial q_1} = 0.$$

(7.13)

It follows that

$$x_1 = q_1 - \frac{1}{2\cos^2 \theta}, \qquad x_2 = \tan \theta.$$

(7.14)

Eliminating θ from (7.14), we get

$$q_1 = x_1 + \frac{1}{2}(x_2^2 + 1) = V(x_1, x_2).$$

(7.15)

When $\kappa \neq 1$, $r = 1/t$, $s = s_0 t^{-4\kappa/(\kappa+1)}$ and H_1 and H_2 are determined from (6.13), we analogously obtain the solution

$$q_1 = \frac{1}{2}\left(\frac{1-\kappa}{1+\kappa} x_1^2 + x_2^2\right) = V(x_1, x_2).$$

(7.16)

The streamlines for the solution (7.1) are determined from the solution of the differential equation

$$\frac{dx_1}{V_1(x_1,\ x_2)} = \frac{dx_2}{V_2(x_1,\ x_2)}. \tag{7.17}$$

For example, when $V(x_1, x_2)$ is given by (7.15) this equation has the form

$$\frac{dx_1}{1} = \frac{dx_2}{x_2}. \tag{7.18}$$

Integrating, we obtain the streamlines

$$x_2 = Ce^{x_1}, \tag{7.19}$$

where C is an arbitrary constant. The picture of the streamlines in the (x_1, x_2) plane is symmetric with respect to the x_1 axis, which itself is a streamline. When $x_1 \to -\infty$ the streamlines asymptotically approach the x_1 axis. There are no intersections of the streamlines within a finite distance of the origin.

When $V(x_1, x_2)$ is given by (7.16), the streamlines are found from the solution of the differential equation

$$\frac{dx_1}{(1-\varkappa)\,x_1} = \frac{dx_2}{(1+\varkappa)\,x_2}. \tag{7.20}$$

The origin is a singular point for equation (7.20). The behavior of the integral curves in a neighborhood of it depends on the value of κ. In particular, when $\kappa < 1$ the origin is a nodal point, all of the streamlines intersect at it, and the coordinate axes are streamlines.

BIBLIOGRAPHY

1. L. D. Landau and E. M. Lifšic, *Mechanics of continuous media,* 2nd ed., GITTL, Moscow, 1953; English transl., parts 1, 2, *A course of theoretical physics,* vols. 6, 7, Pergamon Press, Oxford; Addison-Wesley, Reading, Mass., 1959. MR **16**, 412; **21** #6839.

2. V. A. Sučkov, *Two-dimensional potential flow with stationary streamlines,* Dokl. Akad. Nauk SSSR **165** (1965), 292–293 = Soviet Phys. Dokl. **10** (1965/66), 819–820.

3. N. N. Janenko, *Compatibility theory and integration methods for systems of nonlinear partial differential equations,* Proc. Fourth All-Union Math. Congr. (Leningrad, 1961), vol. II, "Nauka", Leningrad, 1964, pp. 247–252. (Russian) MR **36** #4117.

4. S. P. Finikov, *A course of differential geometry,* GITTL, Moscow, 1952. (Russian) MR **14**, 791.

5. B. Bernstein and T. Y. Thomas, *The differential equations of the stream lines for compressible gas flow,* J. Rational Mech. Anal. **4** (1955), 703–719. MR **17**, 207.

ON THE FIRST DIFFERENTIAL APPROXIMATION METHOD IN THE THEORY OF DIFFERENCE SCHEMES FOR HYPERBOLIC SYSTEMS OF EQUATIONS

UDC 518.332

Ju. I. ŠOKIN

ABSTRACT. The present paper is devoted to an investigation of two of the many problems arising in the theory of difference schemes: the stability of schemes and the approximation viscosity of difference schemes, i.e. the viscosity induced by the structure of the difference schemes.

Bibliography: 23 items.

Introduction

The investigation presented herein is conducted by the method of first differential approximation of difference schemes, the notion of which was introduced in [1] and [2].([1]) This method consists in reducing the investigation of the stability and approximation viscosity of difference schemes to the investigation of a certain system of differential equations (the first differential approximation), which differs from the original system in the case of difference schemes of first order accuracy by terms of order $O(\tau^2)$. The present work continues the investigations begun in [1], [2], [4] and [5].

In §1 we introduce the notion of the first differential approximation of a difference scheme and, for certain classes of schemes of first order accuracy approximating hyperbolic linear systems of differential equations for which a Cauchy problem is posed, we establish a connection between the stability of a difference scheme and the partial parabolicity of its first differential approximation. We consider stability with respect to the L_2 norm. In the one-dimensional case we consider simple difference schemes (with the minimal number of points at the nth level for first order accuracy [1], [2], [6]) and four-point difference schemes (in particular, majorant difference schemes); in the multidimensional case we consider a majorant scheme, Friedrichs' scheme [7], a simple scheme and corresponding splitting schemes. First differential approximations are considered both in parabolic form, when they have the form of the original system of differential equations plus "viscous" terms, and in hyperbolic form. In conclusion we show how the first differential approximation method can be applied to the investigation of the stability of difference schemes approximating nonlinear equations (in particular, the equations of gas dynamics). Although such an extension has not yet received a rigorous basis, practical calculations [8] have demonstrated its validity.

AMS (MOS) subject classifications (1970). Primary 65M05, 35A40; Secondary 35L45.

([1])The first differential approximation algorithm was first used by A. I. Žukov for the elementary equation $\partial u/\partial t = a\partial u/\partial x$ (not published; see [3]).

In §2 we investigate the approximation viscosity of difference schemes. A whole series of investigations (see [3] and [9]–[13], and their bibliographies) has been devoted to the study of their dissipative properties. But here we will be interested in the approximation viscosity of difference schemes from the point of view of preserving contact discontinuities. The investigation is conducted with the use of the first differential approximation. We introduce the K-property of difference schemes, which in the case of the system of equations of gas dynamics is equivalent to the fact that the scheme preserves a contact discontinuity. Necessary and sufficient conditions for possessing the K property are proved for a number of difference schemes. The investigation is conducted for explicit and implicit schemes approximating a hyperbolic system of first-order differential equations with constant coefficients and is then carried over to the case of the equations of gas dynamics in both Lagrangian and Eulerian coordinates.

It should be noted that the first differential approximation method has turned out to be suitable also for investigating the asymptotic behavior of the solutions of difference schemes [14].

The author wishes to take this opportunity to express his deep gratitude to N. N. Janenko, with whom he enjoyed numerous discussions of the main ideas and results of the present work.

§1. Connection between the stability of difference schemes and their first differential approximations

1. We introduce the notion of the first differential approximation of a difference scheme by considering as an example a difference scheme

$$u^{n+1}(x) = \sum_{\alpha} B_{\alpha} u^{n}(x + \tau\lambda_{\alpha}), \qquad (1.1)$$

approximating the following Cauchy problem for a hyperbolic system of first-order differential equations with m unknown functions and s independent space variables:

$$\frac{\partial u}{\partial t} = \sum_{k=1}^{s} A_{k}(x, t) \frac{\partial u}{\partial x_{k}}, \qquad u(x, 0) = u_{0}(x). \qquad (1.2)$$

Here $u = u(x, t)$ is a vector-valued function with m components, $x = \{x_{1}, \ldots, x_{s}\}$ is a point of the s-dimensional real space R_{s}, $0 \leqslant t \leqslant T$, the $A_{k}(x, t)$ and $B_{\alpha}(x, t)$ are real $m \times m$ matrices, the $\lambda_{\alpha} = \{\lambda_{\alpha}^{1}, \ldots, \lambda_{\alpha}^{s}\}$ are displacement vectors, $t = n\tau$ and $u^{n}(x) = u(x, n\tau)$.

For given λ_{α} we say that a difference scheme (1.1) is stable if [3]

$$\|u^{n+1}(x)\| \leqslant [1 + O(\tau)]\|u^{n}(x)\|.$$

The norm here is the L_{2} norm.

By the order of accuracy of a difference scheme (1.1) is meant the largest integer p for which all of the solutions of the system of equations (1.2) having a continuous

$(p + 1)$th derivative satisfy equation (1.1) to within quantities of order $O(\tau^{p+1})$. In the sequel, except when otherwise stipulated, we will confine ourselves to the consideration of difference schemes of first order accuracy.

It is not difficult to show that the following consistency conditions are satisfied if a difference scheme approximates the system of equations (1.2) with first order accuracy:

$$\sum_{\alpha} B_{\alpha} = I, \qquad \sum_{\alpha} \lambda_{\alpha}^{k} B_{\alpha} = A_{k} \qquad (k = 1, \ldots, s), \tag{1.3}$$

where I is the unit matrix.

Expanding the functions $u^{n+1}(x) = u(x, t + \tau)$ and $u^{n}(x + \tau\lambda_{\alpha}) = u(x + \tau\lambda_{\alpha}, t)$ in (1.1) in powers of τ about the point (x, t) and discarding terms of higher than second order in τ, we get

$$\frac{\partial^2 u}{\partial t^2} = \sum_{j, k=1}^{s} P_{jk} \frac{\partial^2 u}{\partial x_j \partial x_k} + N, \tag{1.4}$$

where

$$P_{jk} = \sum_{\alpha} \lambda_{\alpha}^{j} \lambda_{\alpha}^{k} B_{\alpha}, \qquad N = \frac{2}{\tau} \left(\sum_{j=1}^{s} A_j \frac{\partial u}{\partial x_j} - \frac{\partial u}{\partial t} \right).$$

Subtracting from (1.4) the relation

$$\frac{\partial^2 u}{\partial t^2} = \sum_{j, k=1}^{s} A_j A_k \frac{\partial^2 u}{\partial x_j \partial x_k} + \sum_{k=1}^{s} \left(\frac{\partial A_k}{\partial t} + \sum_{j=1}^{s} A_j \frac{\partial A_k}{\partial x_j} \right) \frac{\partial u}{\partial x_k},$$

which is obtained by differentiating (1.2) with respect to t, we get

$$\frac{\partial u}{\partial t} = \sum_{j, k=1}^{s} C_{jk}(x, t) \frac{\partial^2 u}{\partial x_j \partial x_k} + \sum_{j=1}^{s} D_j(x, t) \frac{\partial u}{\partial x_j}, \tag{1.5}$$

where

$$D_j(x, t) = A_j - \frac{\tau}{2} \left(\frac{\partial A_j}{\partial t} + \sum_{k=1}^{s} A_k \frac{\partial A_j}{\partial x_k} \right),$$

$$C_{jk} = \frac{\tau}{2} \left(\sum_{\alpha} \lambda_{\alpha}^{j} \lambda_{\alpha}^{k} B_{\alpha} - A_j A_k \right) = \frac{\tau}{2} (P_{jk} - A_j A_k).$$

The systems of equations (1.4) and (1.5) will be respectively called the hyperbolic and parabolic forms of the first differential approximation of the difference scheme (1.1).

From (1.5) we see that the difference scheme (1.1) appends to the original system of equations (1.2) the "viscous" terms

$$\sum_{j, k=1}^{s} C_{jk} \frac{\partial^2 u}{\partial x_j \partial x_k}.$$

The system of equations (1.5) is said to be partially parabolic at (x^0, t^0) if for all real $\omega_1, \ldots, \omega_s$ the sum of the squares of which is equal to unity the roots of the equation

$$\det \left\| -\sum_{j,\, k=1}^{s} \omega_j \omega_k C_{jk}(x^0,\ t^0) - \lambda I \right\| = 0$$

have nonpositive real parts: $\operatorname{Re} \lambda_j \leqslant 0$ $(j = 1, \ldots, m)$, and strictly parabolic if $\operatorname{Re} \lambda_j < 0$ $(j = 1, \ldots, m)$.

2. Let us consider the case $s = 1$. The system (1.2) takes the form

$$\frac{\partial u}{\partial t} = A(x,\ t)\frac{\partial u}{\partial x}. \tag{1.6}$$

It is assumed that the matrix A has distinct real eigenvalues $\xi_1, \ldots, \xi_m$: $\xi_i \neq \xi_j$ $(i \neq j)$.

THEOREM 1.1. *In order for a simple difference scheme approximating a system of form (1.6) with constant coefficients to be stable, it is necessary that its first differential approximation be partially parabolic. If the matrix A is symmetric, this condition is also sufficient.*

PROOF. In the case under consideration a simple difference scheme has the following form (see [1], [2], [6]) (for both constant and variable coefficients):

$$u^{n+1}(x) = \sum_{\alpha=1}^{2} B_\alpha u^n(x + \tau\lambda_\alpha), \tag{1.7}$$

where

$$\sum_{\alpha=1}^{2} B_\alpha = I, \qquad \sum_{\alpha=1}^{2} \lambda_\alpha B_\alpha = A. \tag{1.8}$$

The parabolic form of the first differential approximation of scheme (1.7) is

$$\frac{\partial u}{\partial t} = D\frac{\partial u}{\partial x} + C\frac{\partial^2 u}{\partial x^2}, \tag{1.9}$$

where

$$D = A - \frac{\tau}{2}\left[\frac{\partial A}{\partial t} + A\frac{\partial A}{\partial x}\right], \qquad C = \frac{\tau}{2}\left[\sum_{\alpha=1}^{2}\lambda_\alpha^2 B_\alpha - A^2\right].$$

If the coefficients of (1.6) are constants, then $D = A$.

We note that in the case under consideration $B_1 B_2 = B_2 B_1$, inasmuch as the consistency conditions (1.8) imply

$$B_1 = (\lambda_1 - \lambda_2)^{-1}(A - \lambda_2 I), \qquad B_2 = (\lambda_1 - \lambda_2)^{-1}(-A + \lambda_1 I).$$

Suppose the difference scheme (1.7) is stable. Then the von Neumann condition is satisfied [9], i.e. the moduli if the eigenvalues $\rho_1, \ldots, \rho_m$ of the amplification matrix of (1.7) do not exceed unity. The amplification matrix of (1.7) is

$$G = B_1 e^{ik\tau\lambda_1} + B_2 e^{ik\tau\lambda_2} = B_1 e^{ik\tau\lambda_1} + (I - B_1) e^{ik\tau\lambda_2}.$$

Hence

$$\rho_j = \mu_j e^{ik\tau\lambda_1} + (1 - \mu_j) e^{ik\tau\lambda_2}$$

and consequently

$$|\rho_j|^2 = 1 - 4\mu_j (1 - \mu_j) \sin^2 \frac{k\tau}{2} (\lambda_1 - \lambda_2) \leqslant 1 \qquad (j = 1, \ldots, m),$$

where the μ_j are the eigenvalues of the matrix B_1. From the latter inequality it follows that $0 \leqslant \mu_j \leqslant 1$, which implies that the eigenvalues of the matrix $B_2 = I - B_1$ are also included between zero and unity. In our case

$$C = \frac{\tau}{2} (\lambda_1 - \lambda_2)^2 B_1 B_2 = \frac{\tau}{2} (\lambda_1 - \lambda_2)^2 B_1 (I - B_1). \qquad (1.10)$$

Consequently the eigenvalues of the matrix C are nonnegative and the system (1.9) is partially parabolic.

Conversely, suppose $A' = A$ (A' is the transpose of A) and the system (1.9) is partially parabolic, i.e. $C \geqslant 0$. Then it follows from (1.10) that $B_\alpha \geqslant 0$ ($\alpha = 1, 2$), and scheme (1.7) is stable since the conditions of Friedrichs' theorem [7] are satisfied.

THEOREM 1.2. *If A is a Lipschitz continuous symmetric matrix, the partial parabolicity of system* (1.9) *is a necessary and sufficient condition for the stability of scheme* (1.7).

PROOF. *Necessity.* As is well known (see [3], [9] or [15]), a necessary condition for the stability of (1.7) is the stability of all of the schemes

$$u^{n+1}(x) = \sum_{\alpha=1}^{2} B_\alpha (x^0, \ t^0) u^n (x + \tau\lambda_\alpha)$$

whose constant coefficients are the values of the coefficients of (1.7) at some point (x^0, t^0). It can be shown in the same way as in the proof of the preceding theorem that at each point (x^0, t^0) the matrices B_1 and B_2 are nonnegative, and therefore (on the basis of (1.10)) $C(x^0, t^0) \geqslant 0$, i.e. the first differential approximation is partially parabolic.

Sufficiency. The partial parabolicity of (1.9) in the case under consideration means that $C \geqslant 0$. From (1.10) and the consistency conditions (1.8) we see that the matrices B_1 and B_2 are nonnegative, symmetric and Lipschitz continuous. It follows according to Friedrichs' theorem that (1.7) is stable. The theorem is proved.

3. Let us approximate (1.6) by a four-point difference scheme

$$u^{n+1}(x) = \sum_{\alpha=-1}^{1} B_\alpha u^n (x + \alpha h), \qquad (1.11)$$

where

$$\sum_{\alpha=-1}^{1} B_\alpha = I, \quad \sum_{\alpha=-1}^{1} \alpha B_\alpha = \varkappa A, \quad \varkappa = \frac{\tau}{h} = \text{const.} \qquad (1.12)$$

The parabolic form of the first differential approximation of (1.11) is still given by (1.9) if D is the same as before and

$$C = \frac{h^2}{2\tau}\left[\sum_{\alpha=-1}^{1} \alpha^2 B_\alpha - \varkappa^2 A^2\right]. \tag{1.13}$$

A difference scheme (1.11) is called a majorant difference scheme (see [1], [2], [16] or [17]) if

$$B_1 = \varkappa A^+, \qquad B_{-1} = -\varkappa A^-, \qquad A^+ \geqslant 0, \qquad A^- \leqslant 0, \qquad A = A^+ + A^-.$$

The matrices A^+ and A^- are chosen so that their nonzero eigenvalues are respectively the positive and negative eigenvalues of A.

THEOREM 1.3. *In order for a majorant scheme* (1.11) *to be stable it is necessary that its first differential approximation be partially parabolic. If $A' = A$, this condition is also sufficient.*

The proof is analogous to the proof of Theorem 1.1. In this case the consistency conditions imply

$$B_0 = I - \varkappa(A^+ - A^-) = I - \varkappa|A|, \qquad C = \frac{h^2}{2\tau}(I - B_0)B_0. \tag{1.14}$$

THEOREM 1.4. *If A is a symmetric matrix and A^+ and A^- are Lipschitz continuous matrices, the partial parabolicity of the first differential approximation of the majorant scheme* (1.11) *is a necessary and sufficient condition for the stability of* (1.11).

PROOF. A necessary condition for the stability of (1.11) is the stability of all of the schemes

$$u^{n+1}(x) = \varkappa A^+(x^0, t^0)u^n(x+h)$$
$$+ [I - \varkappa|A(x^0, t^0)|]u^n(x) - \varkappa A^-(x^0, t^0)u^n(x-h)$$

with constant coefficients. It follows according to Theorem 1.3 that at each point (x^0, t^0) the matrix $C(x^0, t^0) \geqslant 0$, i.e. the first differential approximation of (1.11) is partially parabolic.

Conversely, from the condition $C \geqslant 0$ it follows by virtue of (1.14) that the matrix B_0 is nonnegative, which implies according to Friedrichs' theorem that (1.11) is stable. The theorem is proved.

THEOREM 1.5. *Suppose A is a symmetric matrix. In order for a stable four-point difference scheme* (1.11) *to be a majorant difference scheme it is necessary and sufficient that its coefficients satisfy the conditions*

$$1)\ B'_\alpha = B_\alpha;\ 2)\ B_\alpha B_\beta = B_\beta B_\alpha\ (\alpha,\ \beta = -1,\ 0,\ 1);\ 3)\ B_1 B_{-1} = 0.$$

PROOF. The necessity of the conditions follows from the definition of a majorant scheme and the fact that in this case $A^+A^- = A^-A^+ = 0$ and A^+ and A^- are symmetric matrices. In fact, since A is a symmetric matrix having distinct real eigenvalues, it is orthogonally similar to a diagonal matrix: $a = Q^{-1}AQ$, where a is a diagonal matrix and Q is an orthogonal matrix. Since $A = A^+ + A^-$, where $A^+ \geqslant 0$ and $A^- \leqslant 0$, it follows that $A^+ = Q^{-1}a^+Q$ and $A^- = Q^{-1}a^-Q$, where $a^+ \geqslant 0$ and $a^- \leqslant 0$. Therefore

$$A^+A^- = Q^{-1}a^+QQ^{-1}a^-Q = Q^{-1}a^+a^-Q = Q^{-1}a^-a^+Q = A^-A^+ = 0.$$

Sufficiency. Under our assumptions the amplification matrix G of scheme (1.11) is a normal matrix, which implies

$$GG^* = I - 4(I - B_0)\sin^2\frac{kh}{2}.$$

In view of the stability assumption, we have

$$0 \leqslant B_0 \leqslant I. \tag{1.15}$$

Since the B_α are pairwise commutative symmetric matrices, there exists a single orthogonal matrix Q that reduces each of them to a diagonal matrix: $B_\alpha = Q^{-1}b_\alpha Q$ ($\alpha = -1, 0, 1$), where the b_α are diagonal matrices. From the consistency conditions and (1.15) it follows that $B_1 + B_{-1} = I - B_0 \geqslant 0$, i.e. $b_1 + b_{-1} = I - b_0 \geqslant 0$, and since $b_1b_{-1} = 0$, we consequently get $b_1 \geqslant 0$, $b_{-1} \geqslant 0$ and $B_1 \geqslant 0$, $B_{-1} \geqslant 0$. Since $B_1 - B_{-1} = \kappa A^+$, we get $b_1 - b_{-1} = \kappa a = \kappa a^+ + \kappa a^-$. Therefore $b_1 = \kappa a^+$, $b_{-1} = -\kappa a$, and hence $B_1 = \kappa A^+$, $B_{-1} = -\kappa A^-$, i.e. (1.11) is a majorant scheme.

4. Let us approximate (1.6) by the Friedrichs difference scheme [7]

$$u^{n+1}(x) = \frac{1}{2}[u^n(x+h) + u^n(x-h)] + \frac{1}{2}\kappa A[u^n(x+h) - u^n(x-h)], \tag{1.16}$$

where $\kappa = \tau/h = \text{const}$ and $A = A(x)$.

THEOREM 1.6. *The difference scheme* (1.16) *is stable if its first differential approximation is strictly parabolic.*

PROOF. The parabolic form of the first differential approximation of (1.16) is given by (1.9) if D is the same as before and

$$C = \frac{h^2}{2\tau}(I - \kappa^2 A^2).$$

The strict parabolicity condition means in the case under consideration that $C > 0$, i.e. $\kappa^2 A^2 < I$ and hence $|\kappa\xi_j| < 1$ for all x and $j = 1, \ldots, m$. In particular, $|\kappa\xi_0| < 1$, where $\xi_0 = \max_{j,x}|\xi_j(x)|$. As was shown in [18], the latter inequality is sufficient for the stability of (1.16). The theorem is proved.

5. Let us consider a splitting scheme

$$u^{n+1}(x) = \Omega_s\Omega_{s-1}\ldots\Omega_1 u^n(x) \tag{1.17}$$

approximating the system of equations (1.2).

Scheme (1.17) is equivalent to the scheme

$$u^{n+\frac{j}{s}}(x) = \Omega_j u^{n+\frac{j-1}{s}}(x) \qquad (j=1,\ldots,s), \qquad (1.18)$$

where the $u^{n+j/s}(x)$ $(j=1,\ldots,s-1)$ are auxiliary functions. Thus at the jth fractional level we consider a difference scheme approximating the system of equations

$$\frac{\partial u}{\partial t} = A_j(x,t)\frac{\partial u}{\partial x_j}. \qquad (1.19)$$

Let T_j denote the translation operator with respect to the x_j axis, let $T_{-j} = T_j^{-1}$ and let E denote the identity operator.

THEOREM 1.7. *Suppose* $\Omega_j = \Sigma_{\alpha=1}^2 B_\alpha^j T_j^{\tau\lambda_\alpha^j}$ $(j=1,\ldots,s)$, *i.e. suppose* (1.17) *is a simple splitting scheme. Then scheme* (1.17) *is stable if its first differential approximation is partially parabolic and the* A_j *are Lipschitz continuous symmetric matrices.*

PROOF. The first differential approximation of (1.17) has the form

$$\frac{\partial u}{\partial t} = \sum_{j=1}^{s} D_j \frac{\partial u}{\partial x_j} + \tau \sum_{j>k} A_j \frac{\partial A_k}{\partial x_j}\frac{\partial u}{\partial x_k}$$

$$+ \frac{\tau}{2}\sum_{j=1}^{s}\left[\sum_{\alpha=1}^{2}(\lambda_\alpha^j)^2 B_\alpha^j - A_j^2\right]\frac{\partial^2 u}{\partial x_j^2} + \frac{\tau}{2}\sum_{j>k}(A_jA_k - A_kA_j)\frac{\partial^2 u}{\partial x_j\partial x_k}, \qquad (1.20)$$

in which

$$\sum_{\alpha=1}^{2} B_\alpha^j = I, \qquad \sum_{\alpha=1}^{2}\lambda_\alpha^j B_\alpha^j = A_j \qquad (j=1,\ldots s). \qquad (1.21)$$

The partial parabolicity of (1.20) implies the partial parabolicity of each of the systems

$$\frac{\partial u}{\partial t} = D_j \frac{\partial u}{\partial x_j} + C_j \frac{\partial^2 u}{\partial x_j^2}, \qquad (1.22)$$

where

$$C_j = \frac{\tau}{2}\left[\sum_{\alpha=1}^{2}(\lambda_\alpha^j)^2 B_\alpha^j - A_j^2\right].$$

System (1.22) is the first differential approximation of the difference scheme

$$u^{n+\frac{j}{s}}(x) = \sum_{\alpha=1}^{2} B_\alpha^j T_j^{\tau\lambda_\alpha^j} u^{n+\frac{j-1}{s}}(x)$$

approximating system (1.19). It follows according to Theorem 1.2 that this scheme is stable, i.e.

$$\|\Omega_j\| = 1 + O(\tau) \qquad (j=1,\ldots,s),$$

which implies

$$\|\,\mathcal{Q}_j\,\| = \|\,\mathcal{Q}_s\mathcal{Q}_{s-1}\ldots\mathcal{Q}_1\,\| = 1 + O(\tau).$$

Thus scheme (1.17) is stable under our assumptions. The theorem is proved.

The following assertions are proved analogously.

THEOREM 1.8. *Suppose* (1.17) *is a majorant splitting scheme* [16], [17], *i.e. suppose* $\Omega_j = \kappa_j A_j^+ T_j + (I - \kappa_j |A_j|)E - \kappa_j A_j^- T_{-j}$, *where* $\kappa_j = \tau/h_j = \mathrm{const}$, $A_j = A_j^+ = A_j^-$, $A_j^+ \geqslant 0$, $A_j^- \leqslant 0$, $|A_j| = A_j^+ - A_j^-$. *Then scheme* (1.17) *is stable if its first differential approximation is partially parabolic, the* A_j *are symmetric matrices and the* A_j^+ *and* A_j^- *are Lipschitz continuous matrices.*

THEOREM 1.9. *If the matrix* $\Sigma_1^s \omega_j A_j(x)$ *has mutually distinct eigenvalues* $\mu_1(x, \omega), \ldots, \mu_m(x, \omega)$ *for all real* $\omega = (\omega_1, \ldots, \omega_s)$, $|\omega|^2 = \Sigma_1^s \omega_j^2 = 1$, *a sufficient condition for the stability of scheme* (1.17), *where*

$$\mathcal{Q}_j = \frac{1}{2}(T_j + T_{-j}) + \frac{1}{2}\varkappa A_j(T_j - T_{-j}),$$

is the strict parabolicity of its first differential approximation.

6. Let us approximate the system (1.2) by the majorant scheme

$$u^{n+1}(x) = \left(I - \sum_{j=1}^{s} \varkappa_j |A_j|\right)u^n(x) + \sum_{j=1}^{s}(\varkappa_j A_j^+ T_j - \varkappa_j A_j^- T_{-j})u^n(x). \quad (1.23)$$

The parabolic form of the first differential approximation of this scheme is

$$\frac{\partial u}{\partial t} = \sum_{j=1}^{s} D_j \frac{\partial u}{\partial x_j} - \frac{\tau}{2}\sum_{j\neq k} A_j A_k \frac{\partial^2 u}{\partial x_j \partial x_k} + \sum_{j=1}^{s} \frac{h_j}{2}|A_j|(I - \varkappa_j |A_j|)\frac{\partial^2 u}{\partial x_j^2}, \quad (1.24)$$

while the hyperbolic form is given by (1.4) if N is the same as before and $P_{jk} = \delta_{jk}\kappa_j^{-1}|A_j|$.

In conjunction with (1.2) and (1.4) we will consider the one-dimensional systems of equations

$$\frac{1}{s}\frac{\partial u}{\partial t} = A_j \frac{\partial u}{\partial x_j}, \quad (1.25)$$

$$\frac{1}{s}\frac{\partial^2 u}{\partial t^2} = P_{jj}\frac{\partial^2 u}{\partial x_j^2} \quad (1.26)$$

for $j = 1, \ldots, s$.

Let $[a, b]$ denote the domain of dependence of the point (x, t) for system (1.2), and let $[a^*, b^*]$ denote the domain of dependence of (x, t) for system (1.4).

THEOREM 1.10. *The majorant scheme* (1.23) *is stable if* 1) *the* A_j *are symmetric matrices,* 2) *the* A_j^+ *and* A_j^- *are Lipschitz continuous matrices,* 3) *the system of equations* (1.24) *is partially parabolic, and* 4) $[a, b] \subset [a^*, b^*]$.

PROOF. Condition 4) means that the domain of dependence for system (1.2) is contained in the domain of dependence for system (1.4), which implies that the domain of dependence $[a_j, b_j]$ for system (1.25) is contained in the domain of dependence $[a_j^*, b_j^*]$ for system (1.26). The characteristics of systems (1.25) and (1.26) are the integral curves of the equations

$$\frac{dx_j}{dt} = s\xi^{(j)}, \qquad \frac{dx_j}{dt} = \pm \sqrt{s}\,\mu\left[\left(\frac{1}{\varkappa_j}|A_j|\right)^{1/2}\right],$$

respectively, where $\xi^{(j)}$ is an eigenvalue of the matrix A_j (we have one equation for each eigenvalue) and $\mu[(\kappa_j^{-1}|A_j|)^{1/2}]$ is an eigenvalue of the matrix $(\kappa_j^{-1}|A_j|)^{1/2}$. The inclusion $[a_j, b_j] \subset [a_j^*, b_j^*]$ is then equivalent to the inequality

$$-\sqrt{\frac{s}{\kappa_j}}|\,\xi^{(j)}\,| \leqslant s\xi^{(j)} \leqslant \sqrt{\frac{s}{\varkappa_j}}|\,\xi^{(j)}\,|,$$

i.e. $|\kappa_j| \leqslant 1/s$, and hence

$$\frac{1}{s}I - \varkappa_j|A_j| \geqslant 0 \qquad (j=1,\ldots,s).$$

From the latter inequalities we get that

$$I - \sum_{j=1}^{s} \kappa_j \ A_j| \geqslant 0.$$

Thus the difference scheme (1.23) has Lipschitz continuous nonnegative matrices as coefficients and is consequently stable according to Friedrichs' theorem.

Consider for the system of equations (1.2) the Friedrichs difference scheme [7]

$$u^{n+1}(x) = \sum_{j=1}^{s} \frac{1}{2}\left[\left(\frac{1}{s}I + \varkappa_j A_j\right)T_j + \left(\frac{1}{s}I - \varkappa_j A_j\right)T_{-j}\right]u^n(x). \qquad (1.27)$$

The hyperbolic form of the first differential approximation of this scheme is

$$\frac{\partial^2 u}{\partial t^2} = \sum_{j=1}^{s} \frac{1}{s\varkappa_j^2}\frac{\partial^2 u}{\partial x_j^2} - \frac{2}{\tau}\left(\frac{\partial u}{\partial t} - \sum_{j=1}^{s} A_j\frac{\partial u}{\partial x_j}\right). \qquad (1.28)$$

The domains of dependence of (x, t) for systems (1.2) and (1.28) are respectively denoted by $[a, b]$ and $[a^*, b^*]$. We have

THEOREM 1.11. *The difference scheme* (1.27) *is stable if the A_j are Lipschitz continuous symmetric matrices and* $[a, b] \subset [a^*, b^*]$.

The proof is analogous to the proof of Theorem 1.10.

Let us approximate the system of equations (1.2) by the simple difference scheme (see [1], [2] or [6])

$$u^{n+1}(x) = \sum_{\alpha=1}^{s+1} B_\alpha u^n(x + \tau\lambda_\alpha). \qquad (1.29)$$

In this case the consistency conditons (1.3) are a system of $s + 1$ linear algebraic equations for the $s + 1$ unknowns $B_1, \ldots, B_{s+1}$. Consequently, the matrices B_α are linear combinations of the matrices $I, A_1, \ldots, A_s$.

Let Q, Q' and Q'' denote the domains of dependence of (x, t) for system (1.2), scheme (1.29) and the hyperbolic form of the first differential approximation of (1.29) respectively.

The hyperbolic and parabolic forms of the first differential approximation of (1.29) are (1.4) and (1.5) respcetively, where $\alpha = 1, \ldots, s + 1$.

THEOREM 1.12. *The simple difference scheme* (1.29) *is stable if* 1) *the A_j are Lipschitz continuous symmetric matrices,* 2) *the system of equations* (1.5) ($\alpha = 1, \ldots, s + 1$) *is strictly parabolic,* 3) $Q'' \subset Q'$, *and* 4) *the vectors λ_α do not lie on an $(s - 1)$-dimensional hyperplane.*

PROOF. In order to characterize the domains of dependence Q, Q' and Q'' we introduce their support functions. The support function z of a closed point set Q is the function $z(\xi) = \max_{x \in Q} x\xi$ [19]. The inclusion $Q'' \subset Q'$ then means that $z''(\xi) \leqslant z'(\xi)$. It is not difficult to show under our assumptions that $z(\xi) \leqslant z''(\xi)$ and hence that

$$z(\xi) \leqslant z'(\xi). \tag{1.30}$$

If the λ_α do not lie on an $(s - 1)$-dimensional hyperplane, then, as was shown in [6], (1.30) can be satisfied if the matrices B_α are nonnegative; but then scheme (1.29) is stable. The theorem is proved.

7. The first differential approximation method has also been applied to the investigation of the stability of difference schemes approximating nonlinear equations (in particular, the equations of gas dynamics). Although there does not yet exist a rigorous basis for such an extension, practical calculations have demonstrated its validity [8]. In this case an analysis by the first differential approximation method yields stability restrictions analogous to the restrictions obtained in an analysis by the local Fourier method and permits one to carry out a qualitative comparison of the various stabilities of difference schemes. If system (1.5) is derived by retaining terms of order higher than $O(\tau^2)$ in the expansion of $u^{n+1}(x)$ and replacing them by derivatives in the space variables, additional terms depending on the gradients of the solutions are obtained in the matrices C_{jk} of "viscous terms". These appended terms explain certain difference scheme instabilities that are observed in practice but that are not detectable by the local Fourier method since the latter does not take gradients into account.

An analogous approach has been applied in [20] to the investigation of difference schemes for the nonlinear equations of gas dynamics.

§2. The approximation viscosity of difference schemes

1. The solution of problems of mathematical physics by finite difference methods is affected by the approximation viscosity induced by the structure of the difference schemes. Numerical calculations show that the properties of a difference scheme are es-

pecially strongly revealed by interacting discontinuities. Large deviations of the gas dynamical parameters from their true values, their behavior depending on the dissipative properties of the difference scheme and the difference net, are observed in the interaction domains.

Difference schemes can be divided into two classes in accordance with their dissipative properties: those admitting contact discontinuities (for example, the von Neumann-Richtmyer scheme [21]) and those not admitting contact discontinuities (for example, the Lax scheme [22]). In the first class, as was shown in [3], there exists a family of standing harmonic waves on which the approximative viscosity does not act and which consequently do not decay. To these standing waves there corresponds an "entropy trace", viz. sharp deviations of the entropy and density from their exact values, in contrast to rather small deviations in the pressure and velocity. An entropy trace remains in the neighborhood of a contact boundary and has the character of a strong nonmonotonic deviation from the true density and entropy profiles. In the second class the density and entropy discontinuities are smoothed over. In such schemes the entropy traces vanish and the contact discontinuities are "smeared out" [1], which is not always desirable.

The dissipative properties of a scheme are closely connected with the structure of its first differential approximation. If the parabolic form of a first differential approximation is strictly parabolic, there occurs a strong "smearing out" of discontinuities (not only shock waves but even the interfaces between two media are "smeared out" in the case of the system of equations of gas dynamics). Thus the Lax scheme, which has a strictly parabolic first differential approximation, "smears out" contact boundaries and leads to an artificial diffusion. To avoid this, one should require that the approximative viscosity "not work" on a contact characteristic (the K property).

Below we investigate conditions under which difference schemes possess the K property (see also [4] and [5]).

2. Consider the following difference scheme of first-order accuracy for system (1.6):

$$u^{n+1}(x) = \sum_\alpha B_\alpha u^n (x + \alpha h), \tag{2.1}$$

where the B_α are constant $m \times m$ matrices, $\kappa = \tau/h = \text{const}$ and

$$\sum_\alpha B_\alpha = I, \qquad \sum_\alpha \alpha B_\alpha = \kappa A.$$

The parabolic form of the first differential approximation of (2.1) is

$$\frac{\partial u}{\partial t} = A \frac{\partial u}{\partial x} + C \frac{\partial^2 u}{\partial x^2},$$

where

$$C = \frac{h^2}{2\tau} \left[\sum_\alpha \alpha^2 B_\alpha - \kappa^2 A^2 \right].$$

Suppose the matrix A has mutually distinct real eigenvalues $\xi_1, \ldots, \xi_m$, one of which is equal to zero, and let X denote an arbitrary left eigenvector such that $XA = 0$.

Following [2], we introduce two definitions.

DEFINITION 1. The difference scheme (2.1) has the *K property* if $XC = 0$.

DEFINITION 2. The difference scheme (2.1) has the *strong K property* if

$$XR = X(G - I) = 0,$$

where $G = \Sigma_\alpha B_\alpha e^{i\alpha k h}$ is the amplification matrix of scheme (2.1).

THEOREM 2.1. *In order for the simple difference scheme* (1.7) *to possess the K (strong K) property it is necessary and sufficient that either* $\lambda_1 = 0$, $\lambda_2 \neq 0$ *or* $\lambda_1 \neq 0$, $\lambda_2 = 0$.

PROOF. *Necessity.* Suppose $XC = 0$. From the consistency conditions (1.8) we have

$$C = \frac{\tau}{2}[-A^2 + (\lambda_1 + \lambda_2)A - \lambda_1\lambda_2 I].$$

Therefore

$$XC = -\frac{\tau}{2}\lambda_1\lambda_2 XI = 0$$

and one of the conditions of the theorem must be satisfied.

Suppose $XR = 0$. In the case under consideration

$$R = (\lambda_1 - \lambda_2)^{-1}[(e^{ik\tau\lambda_1} - e^{ik\tau\lambda_2})A + (\lambda_1 e^{ik\tau\lambda_2} - \lambda_2 e^{ik\tau\lambda_1} - \lambda_1 + \lambda_2)I].$$

Therefore

$$XR = (\lambda_1 e^{ik\tau\lambda_2} - \lambda_2 e^{ik\tau\lambda_1} - \lambda_1 + \lambda_2)XI = 0,$$

which implies that

$$\lambda_1 e^{ik\tau\lambda_2} - \lambda_2 e^{ik\tau\lambda_1} - \lambda_1 + \lambda_2 = 0$$

for all k and hence that one of the conditions of the theorem is satisfied.

Sufficiency. If one of the conditions of the theorem is satisfied, then

$$XC = -\frac{\tau}{2}\lambda_1\lambda_2 XI = 0, \qquad XR = 0,$$

i.e. scheme (1.7) has both the K and strong K properties.

COROLLARY 1. *The simple difference scheme* (1.7) *has the K property if and only if it has the strong K property.*

COROLLARY 2. *The Lax scheme* [21] *obtained from scheme* (1.7) *by setting* $\lambda_1 = -\lambda_2 = h/\tau$ *has neither the K nor the strong K property.*

THEOREM 2.2. *The four-point difference scheme* (1.11) *has the K (strong K) property if and only if* $XB_0 = X$.

PROOF. Suppose $XC = 0$. Then

$$X\left(\sum_{\alpha=-1}^{1} \alpha^2 B_\alpha - \varkappa^2 A^2\right) = X\left(B_1 + B_{-1}\right) = 0$$

and the first of the consistency conditions (1.12) implies

$$XI = X\left(B_1 + B_0 + B_{-1}\right) = XB_0.$$

Suppose $XR = 0$. From the consistency conditions we have

$$R = -2\left(I - B_0\right)\sin^2\frac{kh}{2} + i\varkappa A \sin kh$$

and hence $XR = -2(XI - XB_0)\sin^2(kh/2) = 0$, which implies $XB_0 = X$.

Conversely, suppose $XB_0 = X$. From the first of the consistency conditions (1.12) we have $XI = XB_1 + XB_0 + XB_{-1}$, which implies $X(B_1 + B_{-1}) = 0$. Therefore

$$XC = \frac{h^2}{2\tau} X\left[\sum_{\alpha=-1}^{1} \alpha^2 B_\alpha - \varkappa^2 A^2\right] = \frac{h^2}{2\tau} X\left(B_1 + B_{-1}\right) = 0.$$

The theorem is proved.

COROLLARY 1. *The four-point difference scheme* (1.11) *has the strong K property if and only if it has the K property.*

COROLLARY 2. *The majorant scheme* (1.11) *has both the K and strong K properties.*

For when (1.11) is a majorant scheme, $B_0 = I - \kappa(A^+ - A^-)$ (see §1.3), whereas $XA = 0$ implies $XA^+ = XA^- = 0$. Thus $XB_0 = X$ and the condition of Theorem 2.2 is satisfied.

Let us approximate system (1.6) by the predicator-corrector scheme

$$u^*(x) = \sum_\alpha B_\alpha u^n(x + \tau^*\lambda\alpha), \tag{2.2}$$

$$u^{n+1}(x) = u^n(x) + \varkappa A\left[u^*\left(x + \frac{h}{2}\right) - u^*\left(x - \frac{h}{2}\right)\right], \tag{2.3}$$

where (2.2) is an arbitrary difference scheme and (2.3) is a "cross" scheme.

Eliminating the function u^* from (2.2) and (2.3), we obtain the complete scheme

$$u^{n+1}(x) = u^n(x) + \varkappa A \sum_\alpha B_\alpha [u^n(x + \tau^*\lambda_\alpha + h/2) - u^n(x + \tau^*\lambda_\alpha - h/2)].$$

THEOREM 2.3. *The predictor-corrector scheme* (2.2)–(2.3) *has both the K and strong K properties.*

PROOF. It is not difficult to show that in this case $C = (\tau^* - \tau/2)A^2$ and

$$R = \varkappa A \sum 2i\sin\frac{kh}{2} B_\alpha e^{ik\tau^*\lambda\alpha}.$$

Thus $XC = XR = 0$.

COROLLARY. *The Lax-Wendroff scheme* [10] *for system* (1.6) *has both the K and strong K properties.*

Let us consider the following implicit difference scheme for system (1.6):

$$u^{n+1}(x) = u^n(x) + \sum_{\alpha=1}^{2} B_\alpha(\gamma T_0 + \delta E) u^n(x + \tau\lambda_\alpha), \qquad (2.4)$$

where T_0 is the translation operator with respect to the t axis and

$$\sum_{\alpha=1}^{2} B_\alpha = 0, \qquad \sum_{\alpha=1}^{2} \lambda_\alpha B_\alpha = A, \qquad \gamma + \delta = 1, \qquad \gamma, \ \delta \geqslant 0.$$

The matrix C in the first differential approximation of this scheme is

$$C = \frac{\tau}{2}[(\lambda_1 + \lambda_2) A + (2\gamma - 1) A^2],$$

which implies $XC = 0$.

The amplification matrix of (2.4) has the form

$$G = [I - \gamma (\lambda_1 - \lambda_2)^{-1} (e^{ik\tau\lambda_1} - e^{ik\tau\lambda_2}) A]^{-1} \times [I + \delta (\lambda_1 - \lambda_2)^{-1}(e^{ik\tau\lambda_1} - e^{ik\tau\lambda_2}) A].$$

If the eigenvalues ξ_j of A satisfy the inequality

$$|\gamma (\lambda_1 - \lambda_2)^{-1} (e^{ik\tau\lambda_1} - e^{ik\tau\lambda_2}) \xi_j| \leqslant 1 \qquad (j = 1, \ldots, m), \qquad (2.5)$$

then we have the following expansion [23]:

$$G = \sum_{p=0}^{\infty} d_p A^p, \qquad d_p = d_p(\lambda_1, \lambda_2, \gamma, \delta, r).$$

Then

$$R = G - I = \sum_{j=1}^{\infty} d_p A^p,$$

i.e. $XR = 0$. Thus we have

THEOREM 2.4. 1) *The difference scheme* (2.4) *has the K property.*

2) *If the eigenvalues of the matrix A satisfy inequality* (2.5), *then* (2.4) *has the strong K property.*

Let us approximate the system (1.6) by the following difference scheme:

$$u^{n+1}(x) = u^n(x) + \sum_{\alpha=-1}^{1} B_\alpha(\gamma T_0 + \delta E) T^\alpha u^n(x), \qquad (2.6)$$

where

$$\sum_{\alpha=-1}^{1} B_\alpha = 0, \qquad \sum_{\alpha=-1}^{1} \alpha B_\alpha = \varkappa A, \qquad \varkappa = \tau/h = \text{const}, \qquad \gamma + \delta = 1, \qquad \gamma, \ \delta \geqslant 0.$$

THEOREM 2.5. 1) *The difference scheme* (2.6) *has the K property if and only if* $XB_0 = 0$.

2) *If* $XB_0 = 0$ *and the eigenvalues of the matrix* $i\varkappa A \sin kh + B_0(1 - \cos kh)$ *are less than one in absolute value, then* (2.6) *has the strong K property.*

PROOF. 1) In fact, in this case

$$C = \frac{h^2}{2r}\left[-B_0 + \varkappa^2(2\gamma - 1)A^2\right].$$

If $XC = 0$, then it follows from this that $XB_0 = 0$. Conversely, if $XB_0 = 0$, then $XC = 0$.

2) For the amplitude matrix of (2.6) we have the representation

$$G = \left[I - \gamma\sum_{\alpha=-1}^{1} B_\alpha e^{ik\alpha h}\right]^{-1}\left[I + \delta\sum_{\alpha=-1}^{1} B_\alpha e^{ik\alpha h}\right] = [I - \gamma M]^{-1}[I + \delta M],$$

where

$$M = i\varkappa A \sin kh + B_0(1 - \cos kh).$$

If the hypothesis of the theorem holds, then $G = \sum_{p=0}^{\infty}\gamma^p M^p[I + \delta M]$. Hence $R = G - I = \delta M + \sum_{p=1}^{\infty}\gamma^p M^p[I + \delta M]$, and if $XB_0 = 0$, then $XR = 0$. The theorem is proved.

[*Translator's note*. It appears that Theorems 2.4 and 2.5 can be improved by reasoning as follows, starting from the paragraph immediately preceding Theorem 2.4.

The amplification matrix G of (2.4) satisfies the equation

$$(I - \gamma M)G = I + \delta M, \tag{2.5}$$

where

$$M = (\lambda_1 - \lambda_2)^{-1}(e^{ik\tau\lambda_1} - e^{ik\tau\lambda_2})A.$$

Multiplying each side of this equation from the left by X, we see that $XG = X$, which implies $XR = 0$. Thus we have

THEOREM 2.4. *The implicit difference scheme* (2.4) *has both the K and strong K properties.*

Let us approximate (1.6) by the implicit difference scheme

$$u^{n+1}(x) = u^n(x) + \sum_{\alpha=-1}^{1} B_\alpha(\gamma T_0 + \delta E)T^\alpha u^n(x), \tag{2.6}$$

where

$$\sum_{\alpha=-1}^{1} B_\alpha = 0, \qquad \sum_{\alpha=-1}^{1} \alpha B_\alpha = \kappa A, \qquad \kappa = \tau/h = \text{const}, \qquad \gamma + \delta = 1, \qquad \gamma, \delta \geqslant 0.$$

THEOREM 2.5. *The implicit difference scheme* (2.6) *has both the K and strong K properties if and only if* $XB_0 = 0$.

PROOF. In this case we have

$$C = \frac{h^2}{2r}[-B_0 + \kappa^2(2\gamma - 1)A^2].$$

Thus $XC = 0$ if and only if $XB_0 = 0$.

The amplification matrix G of (2.6) satisfies the equation

$$(I - \gamma M)G = I + \delta M,$$

where

$$M = i\kappa A \sin kh + B_0(1 - \cos kh).$$

Multiplying each side of this equation from the left by X, we see that $XG = X$ (and hence $XR = 0$) if and only if $XB_0 = 0$. The theorem is proved.]

3. Let us assume that the matrix A in (1.6) is a constant matrix and has mutually distinct real eigenvalues $\xi_1, \ldots, \xi_m$ none of which is equal to zero. This property is possessed, in particular, by the system of equations of gas dynamics in Eulerian coordinates.

We fix one of the eigenvalues $\xi_j = q$ of A and let X denote an arbitrary left eigenvector such that $XA = qX$.

DEFINITION 3. The difference scheme (2.1) has the *K property* if $XC = 0$.

THEOREM 2.6. *The simple difference scheme* (1.7) *has the K property if and only if either* $\lambda_1 = q$ *or* $\lambda_2 = q$.

PROOF. In the case under consideration $XC = 0$ if and only if
$X[-A^2 + (\lambda_1 + \lambda_2)A - \lambda_1\lambda_2 I] = 0$, i.e.

$$q^2 - (\lambda_1 + \lambda_2)\, q + \lambda_1\lambda_2 = 0 \tag{2.7}$$

and consequently either $\lambda_1 = q$ or $\lambda_2 = q$.

THEOREM 2.7. *The four-point difference scheme* (1.11) *has the K property if and only if* $XB_0 = (1 - \kappa^2 q^2)X$.

PROOF. Here $XC = 0$ if and only if

$$XC = \frac{h^2}{2\tau} X (I - B_0 - \varkappa^2 A^2) = \frac{h^2}{2\tau} (1 - \varkappa^2 q^2) X - \frac{h^2}{2\tau} XB_0 = 0$$

if and only if $XB_0 = (1 - \kappa^2 q^2)X$.

COROLLARY 1. *The Lax scheme obtained from scheme* (1.11) *by setting* $B_0 = 0$ *does not have the K property.*

COROLLARY 2. *The majorant scheme* (1.11) *does not have the K property.*

COROLLARY 3. *The predictor-corrector scheme* (2.2)–(2.3) *does not have the K property.*

We formulate two theorems whose proofs are analogous to the proofs of the preceding theorems.

THEOREM 2.8. *The implicit difference scheme* (2.6) *has the K property if and only if* $XB_0 = (\gamma - \delta)\kappa^2 q^2 X$.

THEOREM 2.9. *The difference scheme*

$$\sum_{\alpha=-1}^{1} D_\alpha u^{n+1} (x + \alpha h) = \sum_{\alpha=-1}^{1} B_\alpha u^n (x + \alpha h),$$

where

$$\sum_{\alpha=-1}^{1} D_\alpha = \sum_{\alpha=-1}^{1} B_\alpha = I, \qquad \sum_{\alpha=-1}^{1} \alpha (B_\alpha - D_\alpha) = \varkappa A,$$

has the K property if and only if

$$X(D_0 - B_0) = 2\varkappa q X(D_1 - D_{-1}) + \varkappa^2 q^2 X.$$

4. The results presented above can be used to construct difference schemes having the K property and approximating the system of equations of gas dynamics. For simplicity we consider the system of equations of gas dynamics in Eulerian coordinates in the case of a polytropic gas, i.e. when the equation of state is $(\gamma - 1)\epsilon\rho = p$:

$$\frac{\partial w}{\partial t} = A \frac{\partial w}{\partial x}, \tag{2.8}$$

where

$$w = \begin{pmatrix} u \\ \rho \\ p \end{pmatrix}, \quad A = \begin{pmatrix} -u & 0 & -\frac{1}{\rho} \\ -\rho & -u & 0 \\ -\gamma p & 0 & -u \end{pmatrix},$$

u is the velocity of the gas, p is the pressure, ρ is the density and ϵ is the specific internal energy.

In this case the matrix A has three eigenvalues: $-u - c$, $-u$ and $-u + c$, where $c^2 = \gamma p/\rho$. The eigenvalue $-u$ has the left eigenvector $X = (0, -c^2, 1)$.

Let us approximate system (2.8) by a four-point difference scheme of first order accuracy:

$$w^{n+1}(x) = \sum_{\alpha=-1}^{1} B_\alpha w^n(x + \alpha h), \tag{2.9}$$

where

$$\sum_{\alpha=-1}^{1} B_\alpha = I, \quad \sum_{\alpha=-1}^{1} \alpha B_\alpha = \varkappa A, \quad B_\alpha = B_\alpha(w), \quad \frac{\tau}{h} = \varkappa = \text{const.}$$

The parabolic form of the first differential approximation of (2.9) is

$$\frac{\partial w}{\partial t} = D \frac{\partial w}{\partial x} + C \frac{\partial^2 w}{\partial x^2},$$

where

$$D = A - \frac{\tau}{2}\left[\frac{\partial A}{\partial t} + A \frac{\partial A}{\partial x}\right], \quad C = \frac{h^2}{2\tau}\left[\sum_{\alpha=-1}^{1} \alpha^2 B_\alpha - \varkappa^2 A^2\right].$$

With the use of the consistency conditions we find that

$$B_1 = {}^1/_2(I + \varkappa A - B_0), \quad B_{-1} = {}^1/_2(I - \varkappa A - B_0),$$
$$C = \frac{h^2}{2\tau}(I - B_0 - \varkappa^2 A^2). \tag{2.10}$$

Let $B_0 = \|b_{ij}^0\|_1^3$. Then

$$C = \frac{h^2}{2\tau}\begin{pmatrix} 1 - b_{11}^0 - \varkappa^2(u^2 + c^2) & -b_{12}^0 & -b_{13}^0 - 2\varkappa^2 \frac{u}{\rho} \\ -b_{21}^0 - 2\rho u\varkappa^2 & 1 - b_{22}^0 - \varkappa^2 u^2 & -b_{23}^0 - \varkappa^2 \\ -b_{31}^0 - 2\gamma p u\varkappa^2 & -b_{32}^0 & 1 - b_{33}^0 - \varkappa^2(u^2 + c^2) \end{pmatrix}.$$

From Theorem 2.7 it follows that the condition $XC = 0$ is equivalent to the equality $XB_0 = (1 - \kappa^2 u^2)X$, which implies

$$b_{31}^0 = c^2 b_{21}^0, \qquad b_{32}^0 = -c^2(1 - b_{22}^0 - \varkappa^2 u^2), \qquad b_{33}^0 = 1 - \varkappa^2 u^2 + c^2 b_{23}^0. \tag{2.11}$$

Thus the four-point difference scheme (2.9) will have the K property if the matrix B_0 in it satisfies conditions (2.11). We obtain a whole family of difference schemes having the K property and depending on the parameters b_{1j}^0 and b_{2j}^0 $(j = 1, 2, 3)$. The first differential approximation method permits us to choose stable schemes from this family by (i) requiring that the first differential approximation be partially parabolic, and (ii) imposing the condition concerning the connection between the domains of dependence of system (2.8) and the hyperbolic form of the first differential approximation, which in this case is

$$\frac{\partial^2 w}{\partial t^2} = Q \frac{\partial^2 w}{\partial x^2} + R,$$

where

$$Q = \frac{h^2}{\tau^2}(I - B_0), \qquad R = -\frac{2}{\tau}\left(\frac{\partial w}{\partial t} - A \frac{\partial w}{\partial x}\right).$$

These requirements lead us to the inequalities

$$\beta_0 \geqslant 0, \qquad \delta \geqslant 0, \tag{2.12}$$

where β_0 and δ are eigenvalues of B_0 and C respectively; in particular,

$$\varkappa^2 u^2 \leqslant 1, \qquad \varkappa^2(u \pm c)^2 \leqslant 1.$$

In addition, conditions (2.12) lead us to the inequalities

$$2 - b_{11}^0 - b_{33}^0 - 2\varkappa^2(u^2 + c^2)$$

$$\pm [(b_{11}^0 - b_{33}^0)^2 + 16\varkappa^4 c^2 u^2 + b_{13}^0 b_{31}^0 + 2\varkappa^2 \frac{u}{\rho} b_{13}^0 + 2\varkappa^2 \gamma p u b_{13}^0]^{1/2} \geqslant 0, \qquad \varkappa^2 u^2 \leqslant 1,$$

$$\pm \sqrt{(b_{11}^0 - b_{33}^0)^2 + 4 b_{13}^0 b_{31}^0} \leqslant b_{11}^0 + b_{33}^0.$$

BIBLIOGRAPHY

1. N. N. Janenko and Ju. I. Šokin, *Correctness of first differential approximations of difference schemes*, Dokl. Akad. Nauk SSSR **182** (1968), 776–778 = Soviet Math. Dokl. **9** (1968), 1215–1217. MR **39** #1138.

2. ———, *On the approximation viscosity of difference schemes*, Dokl. Akad. Nauk SSSR **182** (1968), 280–281 = Soviet Math. Dokl. **9** (1968), 1153–1155. MR **38** #5408.

3. B. L. Roždestvenskiĭ and N. N. Janenko, *Quasilinear systems and their application to the dynamics of gases*, "Nauka", Moscow, 1968; English transl., Transl. Math. Monographs, vol. 30, Amer. Math. Soc., Providence, R. I. (to appear).

4. N. N. Janenko and Ju. I. Šokin, *The first differential approximation of difference schemes for hyperbolic systems of equations*, Sibirsk. Mat. Ž. **10** (1969), 1173–1187 = Siberian Math. J. **10** (1969), 868–880. MR **40** #8287.

5. Ju. I. Šokin, *On the approximation viscosity of implicit difference schemes*, Proc. All-Union Sem. Numer. Methods in Mechanics of Viscous Fluids, vol. 2, "Nauka", Novosibirsk, 1969, pp. 256–261. (Russian) RŽMat. 1970 #4B899,

6. Susan G. Hahn, *Stability criteria for difference schemes*, Comm. Pure Appl. Math. 11 (1958), 243–255. MR **20** #4350.

7. Kurt O. Friedrichs, *Symmetric hyperbolic linear differential equations*, Comm. Pure Appl. Math. 7 (1954), 345–392. MR **16**, 44.

8. N. N. Janenko et al., *On computational methods for problems of gas dynamics with large deformations*, Čisl. Metody Meh. Splošnoĭ Sredy 1 (1970), no. 1, 40 ff.; English transl., Fluid Dynamics Trans. (Proc. Ninth Sympos. Adv. Problems and Methods in Fluid Dynamics, Kazimierz, 1969), Part 1, PWN, Warsaw, 1971, pp. 9–32.

9. R. D. Richtmyer, *Difference methods for initial-value problems*, Interscience Tracts in Pure and Appl. Math., no. 4, Interscience, New York, 1957. MR **20** #438.

10. Peter D. Lax and Burton Wendroff, *Systems of conservation laws*, Comm. Pure Appl. Math. 13 (1960), 217–237. MR **22** #1153.

11. S. K. Godunov, *A difference method for numerical calculation of discontinuous solutions of the equation of hydrodynamics*, Mat. Sb. 47 (89) (1959), 271–306. (Russian) MR **22** #10194.

12. A. A. Samarskiĭ and V. Ja. Arsenin, *On the numerical solution of the equations of gas dynamics with various types of viscosity*, Ž. Vyčisl. Mat. i Mat. Fiz. 1 (1961), 357–360 = USSR Comput. Math. and Math. Phys. 1 (1961), 382–387. MR **24** #B2264.

13. V. F. Kuropatenko, *On difference methods for the equations of hydrodynamics*, Trudy Mat. Inst. Steklov. 74 (1966), 107–137 = Proc. Steklov Inst. Math. 74 (1967), 116–149. MR **35** #3910.

14. Ju. I. Šokin, *The asymptotic behavior of the solutions of difference schemes*, Izv. Sibirsk. Otdel. Akad. Nauk SSSR 1969, no. 3, 65–68. (Russian) MR **41** #2949.

15. Peter D. Lax, *The scope of the energy method*, Bull. Amer. Math. Soc. 66 (1960), 32–35. MR **22** #8222.

16. N. N. Janenko, *Certain questions from the theory of convergence of difference schemes with constant and variable coefficients*, Proc. Fourth All-Union Math. Congr. (Leningrad, 1961), vol. 2, "Nauka", Leningrad, 1964, pp. 613–621. (Russian) MR **36** #4840.

17. N. N. Anučina, *Some difference schemes for hyperbolic systems*, Trudy Mat. Inst. Steklov. 74 (1966), 5–15 = Proc. Steklov Inst. Math. 74 (1967), 1–13. MR **35** #7607.

18. Masaya Yamaguti and Tatsuo Nogi, *An algebra of pseudo difference schemes and its application*, Publ. Res. Inst. Math. Sci. Ser. A 3 (1967/68), 151–166. MR **37** #1762.

19. Peter D. Lax, *Differential equations, difference equations and matrix theory*, Comm. Pure Appl. Math. 11 (1958), 175–194. MR **20** #4572.

20. C. W. Hirt, *Heuristic stability theory for finite-difference equations*, J. Computational Phys. 2 (1967/68), 339–355.

21. John von Neumann and R. D. Richtmyer, *A method for the numerical calculation of hydrodynamic shocks*, J. Appl. Phys. 21 (1950), 232–237. MR **12**, 289.

22. Peter D. Lax, *Weak solutions of nonlinear hyperbolic equations and their numerical computation*, Comm. Pure Appl. Math. 7 (1954), 159–193. MR **16**, 524.

23. F. R. Gantmaher, *The theory of matriccs*, 2nd ed., "Nauka", Moscow, 1966; English tansl. of 1st ed., vols. 1, 2, Chelsea, New York, 1959. MR **21** #6372c; **34** #2585.

Trudy Mat. Inst. Steklov.
122 (1973)

Proc. Steklov Inst. Math.
122 (1973)

ON A GROUP CLASSIFICATION
OF DIFFERENCE SCHEMES FOR
THE SYSTEM OF EQUATIONS OF GAS DYNAMICS

UDC 518.517.951

N. N. JANENKO AND Ju. I. ŠOKIN

ABSTRACT. In this paper conditions are found under which difference schemes admit the same transformations as the approximated system, which in the present case consists of the equations of gas dynamics.

Bibliography: 6 items.

Introduction

The present paper is devoted to a group classification of the difference schemes approximating the equations of gas dynamics. It is known that the equations of gas dynamics are invariant relative to a certain group of point transformations in the space of independent and dependent variables. This invariance is a consequence of the invariance of the conservation laws from which the equations of gas dynamics are derived. Any difference scheme is realized on a concrete net the very presence of which introduces a noninvariance into the computation algorithm. This noninvariance can reveal itself, for example, in the calculations of flow singularities (shock waves, contact boundaries, weak discontinuities) that move at different angles to the mesh lines. The passage to a difference scheme impedes a group analysis, inasmuch as the difference operators do not have the same group properties as the differential operators. It therefore seemed advisable for us to conduct a group classification of difference schemes on the basis of the first differential approximation.

The notion of a first differential approximation was introduced in [1]−[3], and proved to be fruitful for investigating the stability and especially dissipative properties of difference schemes. Being a system of differential equations containing the scheme parameters in their coefficients, the first differential approximation occupies an intermediate position between the original equations of gas dynamics and the difference scheme approximating them, retaining information on the original equations in its hyperbolic part and on the difference scheme in its parabolic part. Here there arises a natural question: To what extent does the first differential approximation preserve the group properties of the equations of gas dynamics? In accordance with this, all schemes can be divided into two classes: those preserving the group properties and those not preserving the group properties.

AMS (MOS) subject classifications (1970). Primary 65P05, 35A35, 35A30.

In this paper we formulate conditions under which the system of equations of the parabolic form of the first differential approximation admits the entire group of transformations [4] admitted by the original system of equations of gas dynamics. In the case of one space variable we consider the system of equations of gas dynamics in both Eulerian and Lagrangian coordinates; in the case of two space variables we consider it in Eulerian coordinates. Also, we distinguish the cases when the system of Lagrangian equations of the first differential approximation satisfies the law of conservation of mass and when it does not satisfy this law. The stability of the classes of schemes we construct is verified by the first differential approximation method.

§1. The case of one space variable

1. Consider the system of equations of gas dynamics in Eulerian coordinates in the case of one space variable,

$$w_t = f_x, \tag{1.1}$$

where

$$w = \begin{pmatrix} \rho u \\ \rho \\ \rho E \end{pmatrix}, \quad f = \begin{pmatrix} -p - \rho u^2 \\ -\rho u \\ -\rho u E - u p \end{pmatrix}, \quad E = \varepsilon + \frac{1}{2} u^2,$$

u is the velocity of the gas, p is the pressure, ρ is the density and ϵ is the specific internal energy. The equation of state of the gas has the form $\epsilon = \epsilon(p, \rho)$.

We approximate system (1.1) by the difference scheme

$$\frac{\Delta_0 w^n(x)}{\tau} = \frac{\Delta_1 + \Delta_{-1}}{2h} f^n(x) + \frac{\Omega\left(x + \frac{h}{2}\right)\frac{\Delta_1}{h} - \Omega\left(x - \frac{h}{2}\right)\frac{\Delta_{-1}}{h}}{h} w^n(x), \tag{1.2}$$

where $t = n\tau$, τ is the mesh width in the direction of the t axis, h is the mesh width in the direction of the x axis, $w^n(x) = w(x, n\tau)$, $\Omega = \|\Omega_{ij}\|_1^3$ is a presently unknown matrix with $\|\Omega\| = O(\tau)$, $\Delta_0 = T_0 - E$, $\Delta_1 = T_1 - E$, $\Delta_{-1} = E - T_{-1}$, T_0 is the translation operator with respect to t, T_1 is the translation operator with respect to x, E is the identity operator and $T_{-1} = T_1^{-1}$. Scheme (1.2) has at least first-order accuracy.

The hyperbolic and parabolic forms of the first differential approximation of (1.2) are respectively

$$\frac{\tau}{2} w_{tt} + w_t = f_x + (\Omega w_x)_x, \tag{1.3}$$

$$w_t = f_x + (C w_x)_x, \tag{1.4}$$

where

$$C = \Omega - \frac{\tau}{2} A^2 = \| \mu_{ij} \|_1^3, \qquad A = \frac{df}{dw},$$

$$\mu_{11} = \Omega_{11} - \frac{\tau}{2}(3u^2 + \theta + z\eta - 3u^2 z),$$

$$\mu_{12} = \Omega_{12} + \frac{\tau}{2} u[2u^2 - 2\theta - 3u^2 z + 3Ez + z\eta],$$

$$\mu_{13} = \Omega_{13} - \frac{3}{2}\tau u z, \qquad \mu_{21} = \Omega_{21} - \frac{\tau}{2} u(2 - z),$$

$$\mu_{22} = \Omega_{22} + \frac{\tau}{2}(u^2 - \theta - u^2 z + Ez),$$

$$\mu_{23} = \Omega_{23} - \frac{\tau}{2} z,$$

$$\mu_{31} = \Omega_{31} - \tau u(E + \eta - u^2 z) - \frac{\tau}{2} u(\theta - Ez),$$

$$\mu_{32} = \Omega_{32} + \frac{\tau}{2}[(2u^2 + Ez)(E + \eta) - (u^2 + E + \eta)\theta - u^2(2u^2 - E) z],$$

$$\mu_{33} = \Omega_{33} - \frac{\tau}{2} u^2(1 + 2z) - \frac{\tau}{2} z(E + \eta),$$

$$z = \frac{p_\varepsilon}{\rho}, \qquad \eta = \frac{p}{\rho}, \qquad \theta = p_\rho, \qquad \Omega_{ij} = \Omega_{ij}(x,\, t,\, w,\, w_x,\, w_t).$$

System (1.4) can be written in the form

$$w_t = f_x + N_x, \tag{1.5}$$

where

$$N = \begin{pmatrix} N_1 \\ N_2 \\ N_3 \end{pmatrix} = \bar{C}\bar{w}_x = C w_x, \qquad \bar{w} = \begin{pmatrix} u \\ \rho \\ p \end{pmatrix}, \qquad \bar{C} = \| \nu_{ij} \|_1^3,$$

$$N_k = \nu_{k1} u_x + \nu_{k2}\rho_x + \nu_{k3}p_x, \qquad \nu_{k1} = \rho(\mu_{k1} + u\mu_{k3}),$$

$$\nu_{k2} = u\mu_{k1} + \mu_{k2} + (E + \rho\varepsilon_\rho)\mu_{k3}, \qquad \nu_{k3} = \rho\varepsilon_p \mu_{k3}.$$

Expressing w and f in terms of $\bar{w}$, we get that (1.5) is equivalent to the system of equations

$$F_1 = u_t + uu_x + \frac{1}{\rho} p_x - \frac{1}{\rho} N_{1x} + \frac{u}{\rho} N_{2x} = 0,$$

$$F_2 = \rho_t + u\rho_x + \rho u_x - N_{2x} = 0, \tag{1.6}$$

$$F_3 = p_t + up_x + a^2 u_x + bN_{2x} + \frac{u}{\rho\varepsilon_p} N_{1x} - \frac{1}{\rho\varepsilon_p} N_{3x} = 0,$$

where

$$a^2 = \frac{p - \rho^2\varepsilon_\rho}{\rho\varepsilon_p}, \qquad b = \frac{\varepsilon + \rho\varepsilon_\rho - {}^1/_2 u^2}{\rho\varepsilon_p}.$$

2. System (1.1) admits the space of operators with basis [4]

$$L_1 = \frac{\partial}{\partial t}, \qquad L_2 = \frac{\partial}{\partial x}, \qquad L_3 = t\frac{\partial}{\partial x} + \frac{\partial}{\partial u}, \qquad L_4 = t\frac{\partial}{\partial t} + x\frac{\partial}{\partial x}, \tag{1.7}$$

to which there respectively correspond the following finite transformations preserving system (1.1): 1) a translation in time, 2) a translation in the space coordinate, 3) a Galilean transformation, and 4) a similarity transformation.

We require that system (1.4) or, equivalently, (1.6) admit the space of operators with basis (1.7), which leads us to certain restrictions on the choice of Ω.

System (1.6) admits the space of operators with basis (1.7) if and only if (see [4])

$$\widetilde{\widetilde{L}}_\alpha F_k\big|_{F_1=0,\; F_2=0,\; F_3=0} = 0, \qquad \alpha = 1,\, 2,\, 3,\, 4;\quad k = 1,\, 2,\, 3,$$

where $\widetilde{\widetilde{L}}_\alpha$ is the operator obtained by successively extending the operator L_α twice. For example, in the case under consideration we have ($\widetilde{L}_\alpha$ is the first extension of L_α)

$$\widetilde{\widetilde{L}}_1 = L_1 = \frac{\partial}{\partial t}, \qquad \widetilde{\widetilde{L}}_2 = L_2 = \frac{\partial}{\partial x},$$

$$\widetilde{L}_3 = t\frac{\partial}{\partial x} + \frac{\partial}{\partial u} - u_x\frac{\partial}{\partial u_t} - \rho_x\frac{\partial}{\partial \rho_t} - p_x\frac{\partial}{\partial p_t},$$

$$\widetilde{L}_4 = t\frac{\partial}{\partial t} + x\frac{\partial}{\partial x} - u_t\frac{\partial}{\partial u_t} - \rho_t\frac{\partial}{\partial \rho_t} - p_t\frac{\partial}{\partial p_t} - u_x\frac{\partial}{\partial u_x} - \rho_x\frac{\partial}{\partial \rho_x} - p_x\frac{\partial}{\partial p_x}.$$

LEMMA. *System (1.6) is invariant relative to a translation in time and a translation in the space coordinate if the elements of the matrix Ω in scheme (1.2) are independent of x and t.*

PROOF. For suppose the condition of the lemma is satisfied. Then

$$\widetilde{\widetilde{L}}_1 F_1 = -\frac{1}{\rho}\frac{\partial}{\partial t}N_{1x} + \frac{u}{\rho}\frac{\partial}{\partial t}N_{2x} = 0,$$

$$\widetilde{\widetilde{L}}_2 F_1 = -\frac{1}{\rho}\frac{\partial}{\partial x}N_{1x} + \frac{u}{\rho}\frac{\partial}{\partial x}N_{2x} = 0,$$

$$\widetilde{\widetilde{L}}_1 F_2 = -\frac{\partial}{\partial t}N_{2x} = 0, \qquad \widetilde{\widetilde{L}}_2 F_2 = -\frac{\partial}{\partial x}N_{2x} = 0,$$

$$\widetilde{\widetilde{L}}_1 F_3 = b\frac{\partial}{\partial t}N_{2x} + \frac{u}{\rho\varepsilon_p}\frac{\partial}{\partial t}N_{1x} - \frac{1}{\rho\varepsilon_p}\frac{\partial}{\partial t}N_{3x} = 0,$$

$$\widetilde{\widetilde{L}}_2 F_3 = b\frac{\partial}{\partial x}N_{2x} + \frac{u}{\rho\varepsilon_p}\frac{\partial}{\partial x}N_{1x} - \frac{1}{\rho\varepsilon_p}\frac{\partial}{\partial x}N_{3x} = 0,$$

since the independence of the Ω_{ij} from x and t implies the independence of the v_{ij} from x and t. The lemma is proved.

THEOREM 1.1. *The system of equations (1.6) of the first differential approximation of the difference scheme (1.2) admits the space of operators with basis (1.7) if the elements of the matrix Ω in (1.2) are independent of x and t and*

$$\frac{\partial}{\partial u_t}\Omega_{ij} = 0, \qquad \frac{\partial}{\partial \rho_t}\Omega_{ij} = 0, \qquad \frac{\partial}{\partial p_t}\Omega_{ij} = 0 \qquad (i,\, j = 1,\, 2,\, 3), \qquad (1.8)$$

$$\frac{\partial}{\partial u}N_{1x} = N_{2x}, \qquad \frac{\partial}{\partial u}N_{2x} = 0, \qquad \frac{\partial}{\partial u}N_{3x} = N_{1x}, \qquad (1.9)$$

$$\left(u_x\frac{\partial}{\partial u_x} + \rho_x\frac{\partial}{\partial \rho_x} + p_x\frac{\partial}{\partial p_x} + 2u_{xx}\frac{\partial}{\partial u_{xx}} + 2\rho_{xx}\frac{\partial}{\partial \rho_{xx}} + 2p_{xx}\frac{\partial}{\partial p_{xx}}\right)N_{\alpha x} = N_{\alpha x}. \qquad (1.10)$$

PROOF. Under our assumptions it follows from the lemma that

$$\tilde{\tilde{L}}_1 F_k = 0, \qquad \tilde{\tilde{L}}_2 F_k = 0 \qquad (k = 1, 2, 3).$$

Further,

$$\tilde{\tilde{L}}_3 F_1 = -\frac{1}{\rho} \, L_3 N_{1x} + \frac{u}{\rho} \, L_3 N_{2x} + \frac{1}{\rho} \, N_{2x} = -\frac{1}{\rho} \left(L_3 N_{1x} - N_{2x} \right)$$

$$+ \frac{u}{\rho} \, L_3 N_{2x} = -\frac{1}{\rho} \left(\frac{\partial}{\partial u} N_{1x} - N_{2x} \right) + \frac{u}{\rho} \frac{\partial}{\partial u} N_{2x} = 0,$$

$$\tilde{\tilde{L}}_3 F_2 = -L_3 N_{2x} = -\frac{\partial}{\partial u} N_{2x} = 0,$$

$$\tilde{\tilde{L}}_3 F_3 = b L_3 N_{2x} - \frac{u}{\rho \varepsilon_p} N_{2x} + \frac{u}{\rho \varepsilon_p} L_3 N_{1x} + \frac{1}{\rho \varepsilon_p} N_{1x} - \frac{1}{\rho \varepsilon_p} L_3 N_{3x}$$

$$= \frac{u}{\rho \varepsilon_p} \left(\frac{\partial}{\partial u} N_{1x} - N_{2x} \right) + \frac{1}{\rho \varepsilon_p} \left(-\frac{\partial}{\partial u} N_{3x} + N_{1x} \right) = 0.$$

Let

$$\mathcal{L}_4 = u_x \frac{\partial}{\partial u_x} + \rho_x \frac{\partial}{\partial \rho_x} + p_x \frac{\partial}{\partial p_x} + 2u_{xx} \frac{\partial}{\partial u_{xx}} + 2\rho_{xx} \frac{\partial}{\partial \rho_{xx}} + 2p_{xx} \frac{\partial}{\partial p_{xx}} .$$

Then

$$\tilde{\tilde{L}}_4 F_2 = -\rho_t - u\rho_x - \rho u_x + \mathcal{L}_4 N_{2x} = -N_{2x} + \mathcal{L}_4 N_{2x} = 0,$$

$$\tilde{\tilde{L}}_4 F_1 = -u_t - u u_x - \frac{1}{\rho} p_x + \frac{1}{\rho} \mathcal{L}_4 N_{1x} - \frac{u}{\rho} \mathcal{L}_4 N_{2x} =$$

$$= -\frac{1}{\rho} N_{1x} + \frac{u}{\rho} N_{2x} + \frac{1}{\rho} \mathcal{L}_4 N_{1x} - \frac{u}{\rho} \mathcal{L}_4 N_{2x} = 0,$$

$$\tilde{\tilde{L}}_4 F_3 = b \left(N_{2x} - \mathcal{L}_4 N_{2x} \right) - \frac{u}{\rho \varepsilon_p} \left(\mathcal{L}_4 N_{1x} - N_{1x} \right) + \frac{1}{\rho \varepsilon_p} \left(\mathcal{L}_4 N_{3x} - N_{3x} \right) = 0.$$

The theorem is proved.

The condition $\partial N_{2x}/\partial u = 0$ means that N_{2x} does not explicitly depend on u. Equalities (1.9) can then be rewritten in the form

$$\frac{\partial}{\partial u} N_{2x} = 0, \qquad N_{1x} = u N_{2x} + R_1, \qquad N_{3x} = \frac{1}{2} u^2 N_{2x} + u R_1 + R_2, \qquad (1.11)$$

$$\frac{\partial}{\partial u} R_1 = 0, \qquad \frac{\partial}{\partial u} R_2 = 0.$$

If

$$\Omega_{21} = \bar{\Omega}_{21} + \frac{\tau}{2} u (2 - z) - u \left(\bar{\Omega}_{23} - \frac{\tau}{2} z \right),$$

$$\Omega_{22} = \bar{\Omega}_{22} - u \left[\bar{\Omega}_{21} - u \left(\Omega_{23} - \frac{\tau}{2} z \right) \right] - \frac{\tau}{4} u^2 (2 - z) - \frac{1}{2} u^2 \left(\Omega_{23} - \frac{\tau}{2} z \right), \qquad (1.12)$$

$$\Omega_{23} = \bar{\Omega}_{23},$$

where

$$\frac{\partial}{\partial u} \bar{\Omega}_{2k} = 0 \qquad (k = 1, 2, 3), \qquad (1.13)$$

then $\partial N_{2x}/\partial u = 0$. Thus Theorem 1.1 remains valid if conditions (1.9) are replaced by conditions (1.11)–(1.13).

3. We require that the law of conservation of mass be satisfied in the first differential approximation of scheme (1.2), i.e. that $N_2 = 0$ and hence that

$$\Omega_{21}=\frac{\tau}{2}\,u\,(2-z), \qquad \Omega_{22}=\frac{\tau}{2}\,(\theta-Ez-u^2+u^2z), \qquad \Omega_{23}=\frac{\tau}{2}\,z. \qquad (1.14)$$

Then system (1.6) takes the form

$$F_1=u_t+uu_x+\frac{1}{\rho}\,p_x-\frac{1}{\rho}\,N_{1x}=0, \qquad F_2=p_t+up_x+pu_x=0,$$

$$F_3=p_t+up_x+a^2u_x+\frac{u}{\rho\varepsilon_p}\,N_{1x}-\frac{1}{\rho\varepsilon_p}\,N_{3x}=0. \qquad (1.15)$$

THEOREM 1.2. *The system of equations (1.15) of the first differential approximation of the difference scheme (1.2) admits the space of operators with basis (1.7), and the law of conservation of mass is satisfied in the first differential approximation, if the elements of the matrix Ω in scheme (1.2) are independent of x and t and the equalities (1.8), (1.10), (1.14) and*

$$\frac{\partial}{\partial u}\,N_{1x}=0, \qquad N_{3x}=(uN_1)_x+R-u_xN_1, \qquad \frac{\partial}{\partial u}\,R=0 \qquad (1.16)$$

are satisfied.

PROOF. From (1.14) we get that $v_{2k}=0$ for $k=1,2,3$, i.e. $N_2=0$, and hence the law of conservation of mass is satisfied in the first differential approximation. The validity of Theorem 1.2 therefore follows from Theorem 1.1, since in this case condition (1.9) takes the form

$$\frac{\partial}{\partial u}\,N_{1x}=0, \qquad \frac{\partial}{\partial u}\,N_{3x}=N_{1x},$$

i.e.

$$N_{3x}=uN_{1x}+R=(uN_1)_x+R-u_xN_1, \qquad \frac{\partial}{\partial u}\,R=0. \qquad (1.17)$$

The independence of N_{1x} from u means that the coefficients v_{1k} ($k=1,2,3$) do not depend on u. If we set

$$\Omega_{11}=\bar{\Omega}_{11}-u\bar{\Omega}_{13}+\frac{3}{2}\,\tau u^2(1-z),$$

$$\Omega_{12}=\bar{\Omega}_{12}-u\bar{\Omega}_{11}+\frac{1}{2}\,u^2\bar{\Omega}_{13}-\frac{\tau}{2}\,u\,[2u^2-3\theta-3u^2z+3Ez], \qquad (1.18)$$

$$\Omega_{13}=\bar{\Omega}_{13}+\frac{3}{2}\,\tau uz, \qquad \frac{\partial}{\partial u}\,\bar{\Omega}_{1k}=0 \qquad (k=1,2,3),$$

we get that

$$v_{11}=\rho\bar{\Omega}_{11}-\frac{\tau}{2}\,\rho\,(\theta+z\eta), \qquad v_{12}=\bar{\Omega}_{12}+(\varepsilon+\rho\varepsilon_p)\,\bar{\Omega}_{13},$$

$$v_{13}=\rho\varepsilon_p\bar{\Omega}_{13}$$

and hence that $\partial N_{1x}/\partial u=0$. Thus condition (1.16) in Theorem 1.2 can be replaced by conditions (1.17)–(1.18).

4. A consequence of system (1.1) is the equation

$$s_t+us_x=0, \qquad (1.19)$$

expressing the conservation of entropy s along a characteristic with slope $dx/dt=u$ (the trajectories). This equation is formally obtained by multiplying system (1.1) from the left by the vector $X=\rho^{-1}(-u,\,u^2-E-\eta,\,1)$, which is a left eigenvector of the matrix A for the eigenvalue $-u$. In this connection we make use of the first law of thermodynamics.

It is known that the presence of a physical viscosity additively entering into p leads to an increase in the entropy, i.e. in place of (1.19) we have

$$s_t + u s_x = \mu u_x^2 = q, \qquad q > 0.$$

If the elements of the matrix Ω in scheme (1.2) are independent of x and t and, in addition, the equalities (1.8), (1.10), (1.14), (1.17), (1.18) and $R = u_x N_1$ are satisfied, then the approximation viscosity additively enters into p in the system of equations of the first differential approximation of (1.2) and leads to an increase in the entropy under the condition $u_x N_1 > 0$.

For in this case system (1.15) can be written in the form

$$(\rho u)_t + (\bar p + \rho u^2)_x = 0, \qquad \rho_t + (\rho u)_x = 0, \qquad (\rho E)_t + (\rho u E + u \bar p)_x = 0,$$

where $\bar p = p + N_1$. Multiplying this system of equations from the left by X, we get

$$s_t + u s_x = \frac{1}{\rho T} u_x N_1 > 0.$$

By virtue of Theorem 1.2, system (1.20) also admits the space of operators with basis (1.7).

We note that if the condition $R = u_x N_1$ is not satisfied, then the approximation viscosity no longer additively enters into p analogously to the physical viscosity, and we have the relation

$$T(s_t + u s_x) = \frac{1}{\rho} R.$$

From conditions (1.9) or, equivalently, (1.16) it follows in particular that if the approximation viscosity enters into the equation of motion ($N_1 \neq 0$), it necessarily also enters into the energy equation ($N_3 \neq 0$). Conversely, if $\partial N_{3x}/\partial u = 0$, then $N_{1x} = 0$ but $N_{3x} \neq 0$, i.e., there exists an approximation viscosity entering into the energy equation ($N_3 \neq 0$) but not the equation of motion ($N_1 = 0$). Difference schemes with an analogous viscosity are used in the numerical solution of a number of problems of gas dynamics.

5. In [2], [3] and [6] we defined the K property for difference schemes approximating hyperbolic systems of equations with constant coefficients. Here we introduce an analogous property for the case under consideration.

DEFINITION. The difference scheme (1.2) has the K *property* if $XC = 0$.

Suppose $N_2 = 0$. Then

$$C = \begin{pmatrix} \mu_{11} & \mu_{12} & \mu_{13} \\ 0 & 0 & 0 \\ \mu_{31} & \mu_{32} & \mu_{33} \end{pmatrix}.$$

The equality $XC = 0$ means that

$$-u\mu_{1k} + \mu_{3k} = 0 \qquad (k = 1,\, 2,\, 3)$$

and hence

$$\Omega_{31} = u\Omega_{11} + \tau u \left(\varepsilon + \eta - u^2 + \frac{1}{2} z\eta + \frac{1}{2} u^2 z - \frac{1}{2} Ez \right),$$

$$\Omega_{32} = u\Omega_{12} + \frac{\tau}{2} u \left[2u^2 - 2\theta - 3u^2 z + 3Ez + z\eta \right] -$$

$$- \frac{\tau}{2} \left[(2u^2 + Ez)(E + \eta) - (u^2 + E)\theta - \eta\theta - u^2(2u^2 - E)z \right],$$

$$\Omega_{33} = u\Omega_{13} + \frac{\tau}{2} z(E + \eta) + \frac{\tau}{2} u^2(1 - z).$$

(1.20)

Thus scheme (1.2) has the K property if $N_2 = 0$ and equalities (1.20) are satisfied. Suppose $N_2 = 0$ and $XC = 0$. Then

$$XCw_x = XN = \frac{1}{\rho}(-uN_1 + N_3) = 0,$$

i.e.

(1.21)

$$N_3 = uN_1$$

and hence

$$(\nu_{31} - u\nu_{11})u_x + (\nu_{32} - u\nu_{12})\rho_x + (\nu_{33} - u\nu_{13})p_x = 0.$$

The latter equality will hold when

$$\mu_{3k} = u\mu_{1k} \qquad (k = 1, 2, 3)$$

and hence, as was shown above, when equalities (1.20) are satisfied. We therefore have

THEOREM 1.3. *The difference scheme* (1.2) *has the K property, the law of conservation of mass is satisfied in the first differential approximation of it and the system of equations of the first differential approximation of it admits the space of operators with basis* (1.7) *if the elements of the matrix* Ω *in it are independent of x and t and the equalities* (1.8), (1.19), (1.14), (1.17), (1.18) *and* (1.21) *are satisfied.*

6. The stability of the family of difference schemes that we have obtained was investigated by the first differential method, i.e. conditions were found under which the first differential approximation is partially parabolic ($C \geqslant 0$) and the domain of dependence for the hyperbolic form of the first differential approximation

$$\frac{\tau}{2} w_{tt} + w_t = \Omega w_{xx} + (A + \Omega_x)w_x$$

is contained in the domain of dependence for the difference scheme. These conditions reduce to the inequalities

$$\frac{\tau}{2} A^2 \leqslant \Omega \leqslant \frac{h^2}{2\tau} I,$$

which provide additional restrictions on the choice of the matrix Ω and of the mesh widths τ and h. It follows in particular that a necessary condition for stability is

$$\frac{\tau^2 u^2}{h^2} \leqslant 1, \qquad \frac{\tau^2(u \pm a)^2}{h^2} \leqslant 1.$$

A group classification of the difference schemes for the system of equations of gas dynamics in Lagrangian coordinates (as well as of the implicit difference schemes) is carried out analogously.

§2. The case of two space variables

1. The system of equations of gas dynamics in the case of two space variables has the form

$$w_t + f_x + g_y = 0, \tag{2.1}$$

where

$$w = \begin{pmatrix} \rho u \\ \rho v \\ \rho \\ \rho E \end{pmatrix}, \quad f = \begin{pmatrix} \rho u^2 + p \\ \rho u v \\ \rho u \\ \rho u E + u p \end{pmatrix}, \quad g = \begin{pmatrix} \rho u v \\ \rho v^2 + p \\ \rho v \\ \rho v E + v p \end{pmatrix},$$

u and v are the components of the velocity vector in the x and y directions respectively, p is the pressure, ρ is the density, $E = \epsilon + \frac{1}{2}(u^2 + v^2)$, and ϵ is the specific internal energy. The equation of state of the gas has the form $p = p(\rho, \epsilon)$.

We approximate system (2.1) by the difference scheme

$$\frac{\Delta_0 w^n(x, y)}{\tau} + \frac{\Delta_1 + \Delta_{-1}}{2h_1} f^n(x, y) + \frac{\Delta_2 + \Delta_{-2}}{2h_2} g^n(x, y)$$
$$= \left\{ \frac{1}{h_1^2} \bar{\Delta}_1 [\Omega_{11}(x, y) \bar{\Delta}_1] + \frac{1}{h_2^2} \bar{\Delta}_2 [\Omega_{22}(x, y) \bar{\Delta}_2] \right. \tag{2.2}$$
$$\left. + \frac{1}{h_1 h_2} \bar{\Delta}_1 [\Omega_{12}(x, y) \bar{\Delta}_2] + \frac{1}{h_1 h_2} \bar{\Delta}_2 [\Omega_{21}(x, y) \bar{\Delta}_1] \right\} w^n(x, y),$$

where h_1 and h_2 are the mesh widths in the x and y directions respectively,

$$\Delta_i = T_i - E, \qquad \Delta_{-i} = E - T_i^{-1},$$
$$\bar{\Delta}_i = T_i^{1/2} - T_i^{-1/2}, \quad T_{-i} = T_i^{-1} \quad (i = 1, 2),$$

T_1 is the translation operator with respect to x, T_2 is the translation operator with respect to y and the Ω_{ij} are presently unspecified matrices with $\| \Omega_{ij} \| = O(\tau)$.

The hyperbolic and parabolic forms of the first differential approximation of scheme (2.2) are respectively

$$\frac{\tau}{2} w_{tt} + w_t + f_x + g_y = \frac{\partial}{\partial x}(\Omega_{11} w_x + \Omega_{12} w_y) + \frac{\partial}{\partial y}(\Omega_{21} w_x + \Omega_{22} w_y), \tag{2.3}$$

$$w_t + f_x + g_y = \frac{\partial}{\partial x}(C_{11} w_x + C_{12} w_y) + \frac{\partial}{\partial y}(C_{21} w_x + C_{22} w_y), \tag{2.4}$$

where

$$C_{ij} = \Omega_{ij} - \frac{\tau}{2} A_i A_j \quad (i, j = 1, 2),$$

$$A_1 = \frac{df}{dw}, \quad A_2 = \frac{dg}{dw}.$$

System (2.4) can be reduced by algebraic transformations to the form

$$F_1 = u_t + uu_x + vu_y + \frac{1}{\rho}\,p_x - \frac{1}{\rho}\,\mathcal{M}_1 + \frac{4}{\rho}\,\mathcal{M}_3 = 0,$$

$$F_2 = v_t + uv_x + vv_y + \frac{1}{\rho}\,p_y - \frac{1}{\rho}\,\mathcal{M}_2 + \frac{v}{\rho}\,\mathcal{M}_3 = 0,$$

$$F_3 = \rho_t + u\rho_x + v\rho_y + \rho u_x + \rho u_y - \mathcal{M}_3 = 0, \tag{2.5}$$

$$F_4 = p_t + up_x + vp_y + a^2 u_x + a^2 v_y + b\mathcal{M}_3$$

$$\qquad + \frac{u}{\rho\varepsilon_p}\,\mathcal{M}_1 + \frac{v}{\rho\varepsilon_p}\,\mathcal{M}_2 - \frac{1}{\rho\varepsilon_p}\,\mathcal{M}_4 = 0.$$

Here

$$a^2 = \frac{p - \rho^2\varepsilon_\rho}{\rho\varepsilon_p}, \qquad b = \frac{E + \rho\varepsilon_\rho - u^2 - v^2}{\rho\varepsilon_p},$$

$$\mathcal{M} = \begin{pmatrix} \mathcal{M}_1 \\ \mathcal{M}_2 \\ \mathcal{M}_3 \\ \mathcal{M}_4 \end{pmatrix} = \begin{pmatrix} N^{(1)}_{1x} + N^{(2)}_{1y} \\ N^{(1)}_{2x} + N^{(2)}_{2y} \\ N^{(1)}_{3x} + N^{(2)}_{3y} \\ N^{(1)}_{4x} + N^{(2)}_{4y} \end{pmatrix} = N^{(1)}_x + N^{(2)}_y,$$

$$N^{(1)} = C_{11}w_x + C_{12}w_y = \| N^{(1)}_k \|^4_1,$$

$$N^{(2)} = C_{21}w_x + C_{22}w_y = \| N^{(2)}_k \|^4_1,$$

$$N^{(i)}_k = \nu^{i1}_{k1}u_x + \nu^{i1}_{k2}v_x + \nu^{i1}_{k3}\rho_x + \nu^{i1}_{k4}p_x + \nu^{i2}_{k1}u_y + \nu^{i2}_{k2}v_y + \nu^{i2}_{k3}\rho_y + \nu^{i2}_{k4}p_y,$$

$$\nu^{ij}_{k1} = \rho\,(\omega^{ij}_{k1} + u\omega^{ij}_{k4}), \qquad \nu^{ij}_{k2} = \rho\,(\omega^{ij}_{k2} + v\omega^{ij}_{k4}),$$

$$\nu^{ij}_{k3} = u\omega^{ij}_{k1} + v\omega^{ij}_{k2} + \omega^{ij}_{k3} + (E + \rho\varepsilon_\rho)\,\omega^{ij}_{k4},$$

$$\nu^{ij}_{k4} = \rho\varepsilon_p\omega^{ij}_{k4}, \qquad C_{ij} = \| \omega^{ij}_{ke} \|^4_1 \qquad (i,\ j = 1,\ 2;\ k,\ e = 1,\ 2,\ 3,\ 4).$$

2. The largest Lie group of point transformations admitted by system (2.1) is of order seven. The corresponding Lie algebra has a basis consisting of the operators [4]

$$L_1 = \frac{\partial}{\partial t}, \qquad L_2 = \frac{\partial}{\partial x}, \qquad L_3 = \frac{\partial}{\partial y},$$

$$L_4 = t\frac{\partial}{\partial x} + \frac{\partial}{\partial u}, \qquad L_5 = t\frac{\partial}{\partial y} + \frac{\partial}{\partial v}, \tag{2.6}$$

$$L_6 = y\frac{\partial}{\partial x} - x\frac{\partial}{\partial y} + v\frac{\partial}{\partial u} - u\frac{\partial}{\partial v},$$

$$L_7 = t\frac{\partial}{\partial t} + x\frac{\partial}{\partial x} + y\frac{\partial}{\partial y},$$

to which there respectively correspond the following finite transformations preserving system (2.1): 1) a translation in time, 2) a translation in the x coordinate, 3) a translation in the y coordinate, 4) a Galilean transformation with respect to the x axis, 5) a Galilean transformation with respect to the y axis, 6) a rotation, and 7) a similarity transformation.

3. Let us find conditions under which system (2.4) or, equivalently, (2.5) admits the transformations given by the operators (2.6). As was shown in [4], this will be true if and only if

$$\tilde{\tilde{L}}_\alpha F_k \big|_{F_1=0,\ F_2=0,\ F_3=0,\ F_4=0} = 0, \qquad \alpha = 1,\ 2,\ 3,\ 4,\ 5,\ 6,\ 7;\quad k = 1,\ 2,\ 3,\ 4.$$

In the case under consideration, for example,

$$\tilde{\tilde{L}}_j = \tilde{L}_j = L_j \qquad (j = 1,\ 2,\ 3),$$

$$\tilde{L}_4 = L_4 - u_x \frac{\partial}{\partial u_t} - v_x \frac{\partial}{\partial v_t} - \rho_x \frac{\partial}{\partial \rho_t} - p_x \frac{\partial}{\partial p_t},$$

$$\tilde{L}_5 = L_5 - u_y \frac{\partial}{\partial u_t} - v_y \frac{\partial}{\partial v_t} - \rho_y \frac{\partial}{\partial \rho_t} - p_y \frac{\partial}{\partial p_t},$$

$$\tilde{L}_6 = L_6 + v_t \frac{\partial}{\partial u_t} - u_t \frac{\partial}{\partial v_t} + (v_x + u_y)\left(\frac{\partial}{\partial u_x} - \frac{\partial}{\partial v_y}\right) + (v_y - u_x)\left(\frac{\partial}{\partial u_y} + \frac{\partial}{\partial v_x}\right)$$

$$+ \rho_y \frac{\partial}{\partial \rho_x} - \rho_x \frac{\partial}{\partial \rho_y} + p_y \frac{\partial}{\partial p_x} - p_x \frac{\partial}{\partial p_y},$$

$$\tilde{L}_7 = L_7 - u_t \frac{\partial}{\partial u_t} - u_x \frac{\partial}{\partial u_x} - u_y \frac{\partial}{\partial u_y} - v_t \frac{\partial}{\partial v_t} - v_x \frac{\partial}{\partial v_x} - v_y \frac{\partial}{\partial v_y} - \rho_t \frac{\partial}{\partial \rho_t}$$

$$- \rho_x \frac{\partial}{\partial \rho_x} - \rho_y \frac{\partial}{\partial \rho_y} - p_t \frac{\partial}{\partial p_t} - p_x \frac{\partial}{\partial p_x} - p_y \frac{\partial}{\partial p_y}.$$

THEOREM 2.1. *If the elements of the matrices Ω_{ij} in the difference scheme* (2.2) *are independent of x, y and t and, in addition,*

$$\frac{\partial}{\partial \varphi_t}\, \Omega_{kl}^{ij} = 0 \qquad (\varphi = u,\ v,\ \rho,\ p),$$

$$\mathcal{M}_1 = u\mathcal{M}_3 + R_1, \qquad \mathcal{M}_2 = v\mathcal{M}_3 + R_2,$$

$$\mathcal{M}_4 = \frac{1}{2}(u^2 + v^2)\mathcal{M}_3 + uR_1 + vR_2 + R,$$

$$\frac{\partial}{\partial u}\mathcal{M}_3 = 0, \qquad \frac{\partial}{\partial u}R_1 = 0, \qquad \frac{\partial}{\partial u}R_2 = 0, \qquad \frac{\partial}{\partial u}R = 0,$$

$$\frac{\partial}{\partial v}\mathcal{M}_3 = 0, \qquad \frac{\partial}{\partial v}R_1 = 0, \qquad \frac{\partial}{\partial v}R_2 = 0, \qquad \frac{\partial}{\partial v}R = 0, \tag{2.7}$$

$$\mathcal{L}_6\mathcal{M}_1 = \mathcal{M}_2 - v\mathcal{M}_3, \qquad \mathcal{L}_6\mathcal{M}_2 = u\mathcal{M}_3 - \mathcal{M}_1,$$

$$\mathcal{L}_6\mathcal{M}_3 = 0, \qquad \mathcal{L}_6\mathcal{M}_4 = -v\mathcal{M}_1 + u\mathcal{M}_2,$$

$$\mathcal{L}_7\mathcal{M}_k = \mathcal{M}_k \qquad (i,\ j = 1,\ 2;\ k,\ l = 1,\ 2,\ 3,\ 4),$$

where

$$\mathcal{L}_6 = (v_x + u_y)\frac{\partial}{\partial u_x} + (v_y - u_x)\frac{\partial}{\partial u_y} + \rho_y \frac{\partial}{\partial \rho_x} - \rho_x \frac{\partial}{\partial \rho_y} +$$

$$\ldots + (v_{xx} + u_{xy} + u_{yx})\frac{\partial}{\partial u_{xx}} + \ldots + (p_{yx} + p_{xy})\frac{\partial}{\partial p_{yy}},$$

$$\mathcal{L}_7 = u_x \frac{\partial}{\partial u_x} + u_y \frac{\partial}{\partial u_y} + \rho_x \frac{\partial}{\partial \rho_x} + \ldots + p_y \frac{\partial}{\partial p_y}$$

$$+ 2u_{xx}\frac{\partial}{\partial u_{xx}} + 2u_{xy}\frac{\partial}{\partial u_{xy}} + \ldots + 2p_{yy}\frac{\partial}{\partial p_{yy}},$$

then the system of equations (2.5) *admits the space of operators with basis* (2.6).

PROOF. Under the fulfilment of the conditions of the theorem we have

$$\tilde{L}_k F_1 = -\frac{1}{\rho}L_k\mathcal{M}_1 + \frac{u}{\rho}L_k\mathcal{M}_3 = 0, \qquad \tilde{L}_k F_2 = -\frac{1}{\rho}L_k\mathcal{M}_2 + \frac{v}{\rho}L_k\mathcal{M}_3 = 0,$$

$$\tilde{L}_k F_3 = -L_k\mathcal{M}_3 = 0,$$

$$\tilde{L}_k F_4 = bL_k\mathcal{M}_3 + \frac{u}{\rho\varepsilon_p}L_k\mathcal{M}_1 + \frac{v}{\rho\varepsilon_p}L_k\mathcal{M}_2 - \frac{1}{\rho\varepsilon_p}L_k\mathcal{M}_4 = 0$$

for $k = 1, 2, 3$. Further,

$$\tilde{\tilde{L}}_4 F_1 = -\frac{1}{\rho}\, \tilde{L}_4 \mathcal{M}_1 + \frac{1}{\rho}\, \tilde{L}_4 (u\mathcal{M}_3) = \frac{v}{\rho}\, \mathcal{M}_3 - \frac{1}{\rho}\, \tilde{L}_4 \mathcal{M}_1$$

$$= \frac{1}{\rho}\, (v\mathcal{M}_3 - L_4 \mathcal{M}_1) = \frac{v}{\rho}\, \mathcal{M}_3 - \frac{1}{\rho}\left(v\,\frac{\partial}{\partial u} - u\,\frac{\partial}{\partial v}\right)\mathcal{M}_3$$

$$= \frac{v}{\rho}\, \mathcal{M}_3 - \frac{1}{\rho}\left(v\,\frac{\partial}{\partial u} - u\,\frac{\partial}{\partial v}\right)(u\mathcal{M}_3 + R_1) = 0,$$

$$\tilde{\tilde{L}}_4 F_2 = -\frac{1}{\rho}\, \tilde{L}_4 \mathcal{M}_2 + \frac{1}{\rho}\, \tilde{L}_4 (v\mathcal{M}_3) = -\frac{u}{\rho}\, \mathcal{M}_3 - \frac{1}{\rho}\, L_4 \mathcal{M}_2$$

$$= \frac{u}{\rho}\, \mathcal{M}_3 - \frac{1}{\rho}\left(v\,\frac{\partial}{\partial u} - u\,\frac{\partial}{\partial v}\right)\mathcal{M}_2 = 0,$$

$$\tilde{\tilde{L}}_4 F_3 = -\tilde{L}_4 \mathcal{M}_3 = -L_4 \mathcal{M}_3 = -\left(v\,\frac{\partial}{\partial u} - u\,\frac{\partial}{\partial v}\right)\mathcal{M}_3 = 0,$$

$$\tilde{\tilde{L}}_4 F_4 = b\tilde{L}_4 \mathcal{M}_3 - \frac{u}{\rho\varepsilon_p}\, \mathcal{M}_3 + \frac{u}{\rho\varepsilon_p}\, \tilde{L}_4 \mathcal{M}_1$$

$$+ \frac{1}{\rho\varepsilon_p}\, \mathcal{M}_1 + \frac{v}{\rho\varepsilon_p}\, \tilde{L}_4 \mathcal{M}_2 - \frac{1}{\rho\varepsilon_p}\, \tilde{L}_4 \mathcal{M}_2 = 0.$$

It can be shown analogously that the conditions of the theorem imply

$$\tilde{\tilde{L}}_\alpha F_k \big|_{F_1=0,\, F_2=0,\, F_3=0,\, F_4=0} = 0 \qquad (\alpha = 5, 6, 7;\ k = 1, 2, 3, 4).$$

The theorem is proved.

4. We will say that the law of conservation of mass is satisfied in the system of equations of the first differential approximation of scheme (2.2) if

$$\mathcal{M}_3 = 0. \tag{2.8}$$

We note that the law of conservation of mass is not satisfied in the system of equations of the first differential approximation of (2.2) if $\Omega_{12} = \Omega_{21} = 0$. In fact, it can be shown in this case that $M_3 \neq 0$ no matter what the choice of the matrices Ω_{11} and Ω_{22}.

It is easily seen that (2.8) holds if

$$\Omega^{11}_{31} = \frac{\tau}{2}\, u\,(2 - z), \qquad \Omega^{11}_{32} = -\frac{\tau}{2}\, vz,$$

$$\Omega^{11}_{33} = \frac{\tau}{2}\,[-u^2 + \theta + (u^2 + v^2 - E)\,z], \qquad \Omega^{11}_{34} = \frac{\tau}{2}\, z,$$

$$\Omega^{12}_{31} = \frac{\tau}{2}\, v, \qquad \Omega^{12}_{32} = \frac{\tau}{2}\, u, \qquad \Omega^{12}_{33} = -\frac{\tau}{2}\, uv, \qquad \Omega^{12}_{34} = 0,$$

$$\Omega^{21}_{31} = \frac{\tau}{2}\, v, \qquad \Omega^{21}_{32} = \frac{\tau}{2}\, u, \qquad \Omega^{21}_{33} = -\frac{\tau}{2}\, uv, \qquad \Omega^{21}_{34} = 0, \tag{2.9}$$

$$\Omega^{22}_{31} = -\frac{\tau}{2}\, uz, \qquad \Omega^{22}_{32} = \frac{\tau}{2}\, v\,(2 - z),$$

$$\Omega^{22}_{33} = \frac{\tau}{2}\,[-v^2 + \theta + (u^2 + v^2 - E)\,z], \qquad \Omega^{22}_{34} = \frac{\tau}{2}\, z,$$

where

$$z = \frac{p_\varepsilon}{\rho}, \qquad \theta = \frac{p}{\rho}, \qquad \Omega_{ij} = \|\Omega^{ij}_{kl}\|^4_1 \qquad (i,\, j = 1, 2),$$

THEOREM 2.2. *The system of equations of the first differential approximation of the difference scheme* (2.2) *admits the space of operators with basis* (2.6), *and the law of conservation of mass is satisfied in it, if the elements of the matrices* Ω_{ij} $(i, j = 1, 2)$ *in* (2.2) *are independent of* x, y *and* t *and the equalities* (2.7), (2.9) *and*

$$\mathcal{M}_4 = u\mathcal{M}_1 + v\mathcal{M}_3 + R, \qquad \frac{\partial}{\partial u}\mathcal{M}_k = 0, \qquad \frac{\partial}{\partial v}\mathcal{M}_k = 0 \qquad (k = 1, 2),$$

$$\frac{\partial}{\partial u}R = 0, \qquad \frac{\partial}{\partial v}R = 0,$$

$$\mathcal{L}_6\mathcal{M}_1 = \mathcal{M}_2, \qquad \mathcal{L}_6\mathcal{M}_2 = -\mathcal{M}_1, \qquad \mathcal{L}_6\mathcal{M}_4 = -v\mathcal{M}_1 + u\mathcal{M}_2,$$
$$\mathcal{L}_7\mathcal{M}_j = \mathcal{M}_j \qquad (j = 1, 2, 3, 4),$$

are satisfied.

The proof is analogous to the proof of Theorem 2.1.

BIBLIOGRAPHY

1. N. N. Janenko and Ju. I. Šokin, *Correctness of first differential approximations of difference schemes*, Dokl. Akad. Nauk SSSR 182 (1968), 776–778 = Soviet Math. Dokl. 9 (1968), 1215–1217. MR 39 #1138.

2. ————, *The first differential approximation of difference schemes for hyperbolic systems of equations*, Sibirsk. Mat. Ž. 10 (1969), 1173–1187 = Siberian Math. J. 10 (1969), 868–880. MR 40 #8287.

3. ————, *First differential approximation method and approximate viscosity of difference schemes*, Phys. Fluids 12 (1969), no. 12, suppl. II, II-28–II-32.

4. L. V. Ovsjannikov, *Group properties of differential equations*, Izdat. Sibirsk. Otdel. Akad. Nauk SSSR, Novosibirsk, 1962. (Russian) MR 26 #264.

5. N. N. Janenko et al., *On computational methods for problems of gas dynamics with large deformations*, Čisl. Metody Meh. Splošnoĭ Sredy 1 (1970), no. 1, 40 ff.; English transl., Fluid Dynamics Trans. (Proc. Ninth Sympos. Adv. Problems and Methods in Fluid Dynamics, Kazimierz, 1969), Part 1, PWN, Warsaw, 1971, pp. 9–32.

6. N. N. Janenko and Ju. I. Šokin, *On the approximation viscosity of difference schemes*, Dokl. Akad. Nauk SSSR 182 (1968), 280–281 = Soviet Math. Dokl. 9 (1968), 1153–1155. MR 38 #5408.